GRAPHING CALCULATOR MANUAL

JUDITH A. PENNA

Indiana University Purdue University India

COLLEGE ALGEBRA: GRAPHS AND MODELS
THIRD EDITION

Marvin L. Bittinger
Indiana University Purdue University Indianapolis

Judith A. Beecher
Indiana University Purdue University Indianapolis

David J. Ellenbogen
Community College of Vermont

Judith A. Penna
Indiana University Purdue University Indianapolis

PEARSON

Addison
Wesley

Boston San Francisco New York
London Toronto Sydney Tokyo Singapore Madrid
Mexico City Munich Paris Cape Town Hong Kong Montreal

Reproduced by Pearson Addison-Wesley from electronic files supplied by the author.

Copyright © 2006 Pearson Education, Inc.
Publishing as Pearson Addison-Wesley, 75 Arlington Street, Boston, MA 02116.

ISBN   0-321-28809-2

1 2 3 4 5 6 BB 08 07 06 05

Contents

The TI-83, TI-83 Plus, and TI-84 Plus Graphics Calculators

Chapter R
Basic Concepts of Algebra

GETTING STARTED

Before turning on the calculator note that there are options above the keys as well as on them. To access the option written on a key, simply press the key. The options written in yellow above the keys are accessed by first pressing the yellow 2nd key in the left column of the keypad and then pressing the key corresponding to the desired option. The green options are accessed by first pressing the green ALPHA key in the left column of the keypad.

Press ON to turn on the calculator. (ON is the key at the bottom left-hand corner of the keypad.) You should see a blinking rectangle, or cursor, on the screen. If you do not see the cursor, try adjusting the display contrast. To do this, first press 2nd . Then press and hold △ to increase the contrast or ▽ to decrease the contrast. If the contrast needs to be adjusted further after the first adjustment, press 2nd again then then hold △ or ▽ to increase or decrease the contrast, respectively.

To turn the calculator off, press 2nd OFF . (OFF is the second operation associated with the ON key.) The calculator will turn itself off automatically after about five minutes without any activity.

Press MODE to display the MODE settings. Initially you should select the settings on the left side of the display. The screen is shown here as it appears on the TI-83 and TI-83 Plus. The mode settings on the TI-84 Plus will appear in a different font and a clock will also appear at the bottom of the screen.

To change a setting on the Mode screen use ▽ or △ to move the cursor to the line of that setting. Then use ▷ or ◁ to move the blinking cursor to the desired setting and press ENTER . Press CLEAR or 2nd QUIT to leave the MODE screen. (QUIT is the second operation associated with the MODE key.) Pressing CLEAR or 2nd QUIT will take you to the home screen where computations are performed.

The TI-83, TI-83 Plus, and TI-84 Plus graphing calculators are very similar in many respects. For that reason, most of the keystrokes and instructions presented in this section of the graphing calculator manual will apply to all three calculators. Where they differ, keystrokes and instructions for using the TI-83 will be given first, followed by those for the TI-83 Plus and the TI-84 Plus.

It will be helpful to read the Getting Started section of the Texas Instruments Guidebook that was packaged with your graphing calculator before proceeding.

ABSOLUTE VALUE

Section R.1, Example 3 Find the distance between -2 and 3.

The distance between -2 and 3 is $|-2-3|$, or $|3-(-2)|$. Absolute value notation is denoted "abs" on the calculator. It is item 1 on the MATH NUM menu.

Note that the $\boxed{(-)}$ key in the bottom row of the keypad must be used to enter a negative number. The blue $\boxed{-}$ key in the right column of the keypad is used for the subtraction operation. Thus, to enter $|-2-3|$ press $\boxed{\text{MATH}}\ \boxed{\triangleright}\ \boxed{\text{ENTER}}$ $\boxed{(-)}\ \boxed{2}\ \boxed{-}\ \boxed{3}\ \boxed{)}\ \boxed{\text{ENTER}}$. To enter $|3-(-2)|$ press $\boxed{\text{MATH}}\ \boxed{\triangleright}\ \boxed{\text{ENTER}}\ \boxed{3}\ \boxed{-}\ \boxed{(}\ \boxed{(-)}\ \boxed{2}\ \boxed{)}\ \boxed{)}\ \boxed{\text{ENTER}}$. Note that the calculator supplies the left parenthesis in the absolute value notation. We close the expression with a right parenthesis. Although it is not necessary to include the inner set of parentheses around -2, they allow the expression to be read more easily so we include them here.

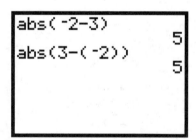

Instead of pressing $\boxed{\text{MATH}}\ \boxed{\triangleright}\ \boxed{\text{ENTER}}$ to access "abs(" and copy it to the home screen, we could have pressed $\boxed{\text{MATH}}\ \boxed{\triangleright}\ \boxed{1}$ since "abs(" is item 1 on the MATH NUM menu. Absolute value notation can also be found as the first item in the CATALOG and copied to the home screen. To do this press $\boxed{\text{2nd}}\ \boxed{\text{CATALOG}}\ \boxed{\text{ENTER}}$. (CATALOG is the second operation associated with the 0 numeric key.)

EDITING ENTRIES

You can recall and edit an entry if necessary. If, for instance, in entering one of the expressions in Example 3 above you pressed 6 instead of 3, first press 2nd ENTRY to return to the last entry. (ENTRY is the second operation associated with the ENTER key.) Then use the ◁ key to move the cursor to 6 and press 3 to overwrite it. If you forgot to type the 2 in the first expression, move the cursor to the subtraction symbol; then press 2nd INS 2 to insert the 2 before that symbol. (INS is the second operation associated with the DEL key.) You can continue to insert symbols immediately after the first insertion without pressing 2nd INS again. If you typed 21 instead of 2, move the cursor to 1 and press DEL. This will delete the 1. If you notice that an entry needs to be edited before you press ENTER to perform the computation, the editing can be done directly without recalling the entry.

The keystrokes 2nd ENTRY can be used repeatedly to recall entries preceding the last one. Pressing 2nd ENTRY twice, for example, will recall the next to last entry. Using these keystrokes a third time recalls the third to last entry and so on. The number of entries that can be recalled depends on the amount of storage they occupy in the calculator's memory.

SCIENTIFIC NOTATION

To enter a number in scientific notation, first type the decimal portion of the number; then press 2nd EE (EE is the second operation associated with the $\cdot$ key.); finally type the exponent, which can be at most two digits. For example, to enter 1.789×10^{-11} in scientific notation, press 1 . 7 8 9 2nd EE (−) 1 1 ENTER. To enter 6.084×10^{23} in scientific notation, press 6 . 0 8 4 2nd EE 2 3 ENTER. The decimal portion of each number appears before a small E while the exponent follows the E.

```
1.789E-11
          1.789E-11
6.084E23
          6.084E23
```

The graphing calculator can be used to perform computations in scientific notation.

Section R.2, Example 8 *Distance to a Star.* The nearest star, Alpha Centauri C, is about 4.22 light-years from Earth. One light-year is the distance that light travels in one year and is about 5.88×10^{12} miles. How many miles is it from Earth to Alpha Centauri? Express your answer in scientific notation.

To solve this problem we find the product $4.22 \times (5.88 \times 10^{12})$. Press 4 . 2 2 × 5 . 8 8 2nd EE 1 2 ENTER. The result is 2.48136×10^{13} miles.

ORDER OF OPERATIONS

Section R.2, Example 9 (b) Calculate: $\dfrac{10 \div (8-6) + 9 \cdot 4}{2^5 + 3^2}$.

In order to divide the entire numerator by the entire denominator, we must enclose both the numerator and the denominator in parentheses. That is, we enter $(10 \div (8-6) + 9 \cdot 4) \div (2^5 + 3^2)$. Press $\boxed{(}$ 1 0 $\boxed{\div}$ $\boxed{(}$ 8 $\boxed{-}$ 6 $\boxed{)}$ $\boxed{+}$ 9 $\boxed{\times}$ 4 $\boxed{)}$ $\boxed{\div}$ $\boxed{(}$ 2 $\boxed{\wedge}$ 5 $\boxed{+}$ 3 $\boxed{x^2}$ $\boxed{)}$ $\boxed{\text{ENTER}}$. Note that 3^2 can be entered either as 3 $\boxed{x^2}$ or as 3 $\boxed{\wedge}$ 2.

RADICAL NOTATION

We can use the square-root, cube-root, and xth-root features to simplify radical expressions.

Section R.6, Example 1 Simplify each of the following.

a) $\sqrt{36}$ b) $-\sqrt{36}$ c) $\sqrt[5]{\dfrac{32}{243}}$ d) $\sqrt[3]{-8}$ e) $\sqrt[4]{-16}$

a) To find $\sqrt{36}$, press $\boxed{\text{2nd}}$ $\boxed{\sqrt{}}$ 3 6 $\boxed{)}$ $\boxed{\text{ENTER}}$. ($\sqrt{}$ is the second operation associated with the $\boxed{x^2}$ key.) Note that the calculator supplies a left parenthesis with the radical symbol and we close the expression with a right parenthesis.

b) To find $-\sqrt{36}$, press $\boxed{(-)}$ $\boxed{\text{2nd}}$ $\boxed{\sqrt{}}$ 3 6 $\boxed{)}$ $\boxed{\text{ENTER}}$.

c) We will use the xth-root feature from the MATH menu to find $\sqrt[5]{\dfrac{32}{243}}$. We will also use ▷Frac to express the result as a fraction. Press 5 | MATH | 5 | (| 3 2 | ÷ | 2 4 3 |) | | MATH | 1 | ENTER |. The first 5 is the index of the radical, and the second 5 is used to select item 5, the xth-root, from the MATH menu.

d) We will use the cube-root feature from the MATH menu to find $\sqrt[3]{-8}$. Press | MATH | 4 | (−) | 8 |) | | ENTER |.

e) To enter $\sqrt[4]{-16}$ press 4 | MATH | 5 | (−) | 1 6 | ENTER |. When the calculator is set in REAL mode, we get an error message indicating that the answer is nonreal.

RATIONAL EXPONENTS

We can add and subtract rational exponents on a graphing calculator.

Section R.6, Example 9 (a) Simplify and then, if appropriate, write radical notation for $x^{5/6} \cdot x^{2/3}$.

To find this product we first add the exponents, expressing the sum as a fraction. Press 5 | ÷ | 6 | + | 2 | ÷ | 3 | MATH | 1 | ENTER |. These keystrokes tell the calculator to add $\dfrac{5}{6}$ and $\dfrac{2}{3}$; then they access the MATH submenu of the MATH menu, copy item 1 "▷Frac" to the home screen, and display the result as a fraction.

Then, proceeding as on page 45 of the text, we find that the final result in radical notation is $x\sqrt{x}$.

Chapter 1
Graphs, Functions, and Models

GRAPHING EQUATIONS

Section 1.1, Example 4 Graph $3x - 5y = -10$.

First we solve for y as shown in the text, obtaining $y = \dfrac{3}{5}x + 2$. Then press $\boxed{Y =}$ to access the equation-editor screen. Clear any equations that are present. To clear an entry for Y_1, for example, position the cursor beside " $Y_1 =$" and press $\boxed{\text{CLEAR}}$. Do this for each existing entry. Also turn off any plots that are turned on. If the name of a plot is highlighted at the top of the equation-editor screen, it is turned on. To turn it off, position the cursor over it and press $\boxed{\text{ENTER}}$. The name will no longer be highlighted, indicating that it is now turned off.

Next enter the equation by positioning the cursor beside "$Y_1 =$" and pressing $\boxed{(}$ $\boxed{3}$ $\boxed{\div}$ $\boxed{5}$ $\boxed{)}$ $\boxed{\text{X, T, }\Theta, n}$ $\boxed{+}$ 2. Although the parentheses are not necessary, the equation is more easily read when they are used.

```
Plot1 Plot2 Plot3
\Y1 ■(3/5)X+2■
\Y2=
\Y3=
\Y4=
\Y5=
\Y6=
\Y7=
```

Next we set the viewing window. This is the portion of the coordinate plane that appears on the calculator's screen. It is defined by the minimum and maximum values of x and y: Xmin, Xmax, Ymin, and Ymax. The notation [Xmin, Xmax, Ymin, Ymax] is used to represent these window settings or dimensions. For example, $[-12,\ 12,\ -8,\ 8]$ denotes a window that displays the portion of the x-axis from -12 to 12 and the portion of the y-axis from -8 to 8. In addition, the distance between tick marks on the axes is defined by the settings Xscl and Yscl. In this manual Xscl and Yscl will be assumed to be 1 unless noted otherwise. The Xres setting sets the pixel resolution. We usually select Xres = 1. The window corresponding to the settings $[-20,\ 30,\ -12,\ 20]$, Xscl = 5, Yscl = 2, Xres = 1, is shown below.

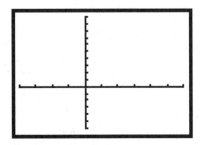

Press the WINDOW key on the top row of the keypad to display the current window settings on your graphing calculator. The standard settings are shown below.

```
WINDOW
 Xmin=-10
 Xmax=10
 Xscl=1
 Ymin=-10
 Ymax=10
 Yscl=1
 Xres=1
```

To change a setting, position the cursor beside the setting you wish to change and enter the new value. For example, to change from the standard settings to $[-20, 30, -12, 20]$, Xscl = 5, Yscl = 2, on the WINDOW screen press (−) 2 0 ENTER 3 0 ENTER 5 ENTER (−) 1 2 ENTER 2 0 ENTER 2 ENTER. Recall that the (−) key in the bottom row of the keypad must be used to enter a negative number. The blue − key in the right column of the keypad is used for the subtraction operation. The ▽ key may be used instead of ENTER after typing each window setting. To see the window shown on the previous page, press the GRAPH key on the top row of the keypad.

QUICK TIP: To return quickly to the standard window setting $[-10, 10, -10, 10]$, Xscl = 1, Yscl = 1, press ZOOM 6.

The standard window is a good choice for the graph of the equation $y = \frac{3}{5}x + 2$. Either enter these dimensions in the WINDOW screen and then press GRAPH to see the graph or simply press ZOOM 6 to select the standard window and see the graph.

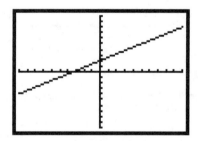

THE TABLE FEATURE

A table of x-and y-values representing ordered pairs that are solutions of an equation can be displayed on a graphing calculator.

Section 1.1, Example 5 Graph: $y = x^2 - 9x - 12$.

If a hand-drawn graph of an equation is desired, a graphing calculator can be used to create a table of ordered pairs that are solutions of the equation. These ordered pairs can then be plotted and connected with a smooth curve to graph the equation. To create such a table for the equation $y = x^2 - 9x - 12$, we first enter the equation on the equation-editor screen. Then press 2nd TblSet to display the table set-up screen. (TblSet is the second function associated with the

WINDOW key.) You can choose to supply the x-values yourself or you can set the graphing calculator to supply them. To have the graphing calculator supply the x-values, set "Indpnt" to "Auto" by positioning the cursor over "Auto" and pressing ENTER. "Depend" should also be set to "Auto."

When "Indpnt" is set to "Auto," the graphing calculator will supply values for x, beginning with the value specified as TblStart and continuing by adding the value of ΔTbl to the preceding value for x. We will display a table of values that starts with $x = -3$ and adds 1 to the preceding x-value. Press (−) 3 ▽ 1 or (−) 3 ENTER 1 to select a minimum x-value of −3 and an increment of 1. To display the table press 2nd TABLE. (TABLE is the second operation associated with the GRAPH key.) We can use the △ and ▽ keys to scroll up and down in the table to find values other than those shown here.

GRAPHING CIRCLES

If the center and radius of a circle are known, the circle can be graphed using the Circle feature from the DRAW menu.

Section 1.1, Example 11 Graph $(x + 5)^2 + (y - 2)^2 = 16$.

The center of this circle is $(-5, 2)$ and its radius is 4. To graph it using the Circle feature from the DRAW menu first press Y = and clear all previously entered equations. Then select a square window. (See page 69 of the text for a discussion on squaring the viewing window.) We will use $[-12, 3, -3, 7]$. Press 2nd QUIT to go to the home screen. Then press 2nd DRAW 9 to display "Circle(." Enter the coordinates of the center and the radius, separating the entries by commas, and close the parentheses: (−) 5 , 2 , 4) ENTER.

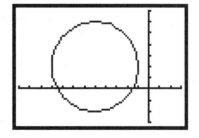

FINDING FUNCTION VALUES

When a formula for a function is given, function values can be found in several ways.

Section 1.2, Example 4 (b) For $f(x) = 2x^2 - x + 3$, find $f(-7)$.

Method 1: Substitute the inputs directly in the formula. Press 2 $\boxed{(}$ $\boxed{(-)}$ 7 $\boxed{)}$ $\boxed{x^2}$ $\boxed{-}$ $\boxed{(}$ $\boxed{(-)}$ 7 $\boxed{)}$ $\boxed{+}$ 3 $\boxed{\text{ENTER}}$. Although it is not necessary to use the second set of parentheses, they allow the expression to be read more easily so we include them here.

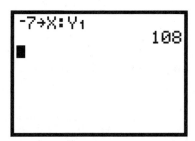

Method 2: Enter $y_1 = 2x^2 - x + 3$ on the "Y =" screen. Then press $\boxed{\text{2nd}}$ $\boxed{\text{QUIT}}$ to go to the home screen. To find $f(-7)$, the value of y_1 when $x = -7$, press $\boxed{(-)}$ 7 $\boxed{\text{STO} \triangleright}$ $\boxed{\text{X, T, } \Theta, n}$ $\boxed{\text{ALPHA}}$ $\boxed{:}$ $\boxed{\text{VARS}}$ $\boxed{\triangleright}$ 1 1 $\boxed{\text{ENTER}}$. (: is the ALPHA operation associated with the $\boxed{.}$ key.) This series of keystrokes stores -7 as the value of x and then substitutes it in the function y_1.

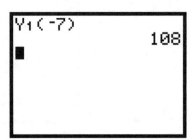

Method 3: Enter $y_1 = 2x^2 - x + 3$ on the "Y =" screen and press $\boxed{\text{2nd}}$ $\boxed{\text{QUIT}}$ to go to the home screen. To find $f(-7)$, press $\boxed{\text{VARS}}$ $\boxed{\triangleright}$ 1 1 $\boxed{(}$ $\boxed{(-)}$ 7 $\boxed{)}$ $\boxed{\text{ENTER}}$. Note that this entry closely resembles function notation.

Method 4: The TABLE feature can also be used to find function values. Enter $y_1 = 2x^2 - x + 3$ on the "Y =" screen. Then set up a table Ask mode, by pressing $\boxed{\text{2nd}}$ $\boxed{\text{TBLSET}}$, moving the cursor over "Indpnt: Ask," and pressing $\boxed{\text{ENTER}}$. In Ask mode, you supply x-values and the calculator returns the corresponding y-values. The settings for TblStart and

ΔTbl are irrelevant in this mode. Press $\boxed{\text{2nd}}$ $\boxed{\text{TABLE}}$ to display the TABLE screen. Then press $\boxed{(-)}$ 7 $\boxed{\text{ENTER}}$ to find $f(-7)$.

Method 5: We can also use the Value feature from the CALC menu to find $f(-7)$. To do this, graph $y_1 = 2x^2 - x + 3$ in a window that includes the x-value -7. We will use the standard window. Then press $\boxed{\text{2nd}}$ $\boxed{\text{CALC}}$ 1 or $\boxed{\text{2nd}}$ $\boxed{\text{CALC}}$ $\boxed{\text{ENTER}}$ to access the CALC menu and select item 1, Value. Now supply the desired x-value by pressing $\boxed{(-)}$ 7. Press $\boxed{\text{ENTER}}$ to see X $= -7$, Y $= 108$ at the bottom of the screen, Thus, $f(-7) = 108$.

GRAPHS OF FUNCTIONS

The TI-83, TI-83 Plus, and TI-84 Plus graphing calculators do not use function notation. To graph a function, first replace the function notation with y. For example, to graph $f(x) = x^2 - 5$ replace $f(x)$ with y. Then enter the equation $y = x^2 - 5$ on the equation-editor screen and graph it as described on page 9 of this manual.

LINEAR REGRESSION

We can use the Linear Regression feature in the STAT CALC menu to fit a linear equation to a set of data.

Section 1.4, Example 10 The table on the following page shows the number of U. S. travelers to foreign countries, in millions, for years since 1994. Fit a regression line to the data. Then use the function to estimate the number of U. S. travelers to foreign countries in 2006.

Years, x	Number of U. S. Travelers to Foreign Countries (in millions)
1994, 0	4.65
1995, 1	5.08
1996, 2	5.23
1997, 3	5.29
1998, 4	5.63
1999, 5	5.75
2000, 6	6.08

We will enter the data as ordered pairs on the STAT list editor screen. To clear any existing lists press $\boxed{\text{STAT}}$ 4 $\boxed{\text{2nd}}$ $\boxed{\text{L}_1}$, $\boxed{\text{2nd}}$ $\boxed{\text{L}_2}$, $\boxed{\text{2nd}}$ $\boxed{\text{L}_3}$, $\boxed{\text{2nd}}$ $\boxed{\text{L}_4}$, $\boxed{\text{2nd}}$ $\boxed{\text{L}_5}$, $\boxed{\text{2nd}}$ $\boxed{\text{L}_6}$ $\boxed{\text{ENTER}}$. (L_1 through L_6 are the second operations associated with the numeric keys 1 through 6.) The lists can also be cleared by first accessing the STAT list editor screen by pressing $\boxed{\text{STAT}}$ $\boxed{\text{ENTER}}$ or $\boxed{\text{STAT}}$ 1. These keystrokes display the STAT EDIT menu and then select the Edit option from that menu. Then, for each list that contains entries, use the arrow keys to move the cursor to highlight the name of the list at the top of the column and press $\boxed{\text{CLEAR}}$ $\boxed{\triangledown}$ or $\boxed{\text{CLEAR}}$ $\boxed{\text{ENTER}}$.

Once the lists are cleared, we can enter the coordinates of the points. We will enter the first coordinates in L_1 as the number of years since 1994 and the second coordinates in L_2. Position the cursor at the top of column L_1, below the L_1 heading. To enter 0 press 0 $\boxed{\text{ENTER}}$. Continue typing the first coordinates, 1, 2, 3, 4, 5, and 6 in order, each followed by $\boxed{\text{ENTER}}$. The entries can be followed by $\boxed{\triangledown}$ rather than $\boxed{\text{ENTER}}$ if desired. Press $\boxed{\triangleright}$ to move to the top of column L_2. Type the second coordinates, 4.65, 5.08, 5.23, 5.29, 5.63, 5.75, and 6.08, in succession, each followed by $\boxed{\text{ENTER}}$ or $\boxed{\triangledown}$. Note that the coordinates of each point must be in the same position in both lists.

The graphing calculator can then plot these points, creating a scatterplot of the data. To do this we turn on the STAT PLOT feature. To access the STAT PLOT screen, press $\boxed{\text{2nd}}$ $\boxed{\text{STAT PLOT}}$. (STAT PLOT is the second operation associated with the $\boxed{\text{Y} =}$ key in the upper left-hand corner of the keypad.) We will use Plot 1. Access it by highlighting 1 and pressing $\boxed{\text{ENTER}}$ or simply by pressing 1. Now position the cursor over On and press $\boxed{\text{ENTER}}$ to turn on Plot 1. The entries Type, Xlist, and Ylist should be as shown below. The last item, Mark, allows us to choose a box, a cross, or a dot for each point. Here we have selected a box. To select Type and Mark, position the cursor over the appropriate selection and press $\boxed{\text{ENTER}}$. Use the L_1 and L_2 keys (associated with the 1 and 2 numeric keys) to select Xlist and Ylist. The entries should be as shown below.

If the plot has previously been set up as desired, it can be turned on from the equation-editor, or "Y =", screen. Press $\boxed{Y =}$, the key at the top left-hand corner of the keypad, to go to this screen. Then, assuming Plot 1 has not yet been turned on, position the cursor over Plot 1 and press $\boxed{\text{ENTER}}$. Plot 1 will now be highlighted.

Note that there should be no equations entered on the equation-editor screen. If there are equations present, clear them as described on page 9 of this manual. If this is not done, the equations that are currently entered will be graphed along with the data points that are entered.

Now we select a window. The x-values range from 0 through 6 and the y-values from 4.65 through 6.08, so one good choice might be $[-2, 7, 0, 7]$. We can enter these dimensions and then press $\boxed{\text{GRAPH}}$ to see the scatterplot. Instead of entering the window dimensions directly, we can press $\boxed{\text{ZOOM}}$ 9 after entering the coordinates of the points in lists, turning on Plot 1, and selecting Type, Xlist, Ylist, and Mark. This activates the ZoomStat operation which automatically defines a viewing window that displays all the points. We use $\boxed{\text{ZOOM}}$ 9 to produce the scatterplot shown below.

The calculator's linear regression feature can be used to fit a linear equation to the data. Once the data have been entered in the lists, press $\boxed{\text{STAT}}$ $\boxed{\triangleright}$ 4 $\boxed{\text{ENTER}}$ to select LinReg($ax + b$) from the STAT CALC menu and to display the coefficients a and b of the regression equation $y = ax + b$.

If the diagnostics have been turned on, values for r^2 and r will be displayed. These numbers indicate how well the regression line fits the data.

If you wish to select DiagnosticOn mode, press $\boxed{\text{2nd}}$ $\boxed{\text{CATALOG}}$ and use $\boxed{\triangledown}$ to position the triangular selection cursor beside DiagnosticOn. To alleviate the tedium of scrolling through many items to reach DiagnosticOn, press $\boxed{\text{D}}$ after pressing $\boxed{\text{2nd}}$ $\boxed{\text{CATALOG}}$ to move quickly to the first catalog item that begins with the letter D. (D is the ALPHA operation associated with the $\boxed{x^{-1}}$ key.) Note that it is not necessary to press $\boxed{\text{ALPHA}}$ before $\boxed{\text{D}}$ when the catalog is displayed. Then use $\boxed{\triangledown}$ to scroll to DiagnosticOn. Press $\boxed{\text{ENTER}}$ to paste this instruction to the home screen and then press $\boxed{\text{ENTER}}$ a second time to set the mode. To select DiagnosticOff mode, press $\boxed{\text{2nd}}$ $\boxed{\text{CATALOG}}$, position the selection cursor beside DiagnosticOff, press $\boxed{\text{ENTER}}$ to paste this instruction to the home screen, and then press $\boxed{\text{ENTER}}$ again to set this mode.

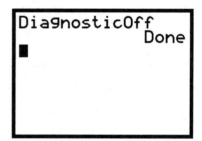

Immediately after the regression equation is found it can be copied to the equation-editor screen. To copy it as Y_1, for example, first note that any previous entry in Y_1 must be cleared first. Press $\boxed{\text{Y =}}$ and position the cursor beside Y_1. Then press $\boxed{\text{VARS}}$ 5 $\boxed{\triangleright}$ $\boxed{\triangleright}$ 1. These keystrokes select Statistics from the VARS menu, then select the EQ (Equation) submenu, and finally select the RegEq (Regression Equation) from this submenu.

Before the regression equation is found, it is possible to select a y-variable to which it will be stored on the equation editor screen. After the data have been stored in the lists and the equation previously entered as Y_1 has been cleared, press

STAT $\triangleright$ 4 VARS $\triangleright$ 1 1 ENTER . The coefficients of the regression equation will be displayed on the home screen, and the regression equation will also be stored as Y_1 on the equation-editor screen.

Once the regression equation has been copied to the equation-editor screen, we can see the graph of the equation along with the scatterplot by pressing GRAPH .

To estimate the number of U. S. travelers to foreign countries in 2006, evaluate the regression equation for $x = 12$. (2006 is 12 years after 1994.) Use any of the methods for evaluating a function presented earlier in this chapter. (See pages 12 and 13 of this manual.) We will use function notation on the home screen. When $x = 12, y \approx 7.33$, so we estimate that there will be about 7.33 million U. S. travelers to foreign countries in 2006.

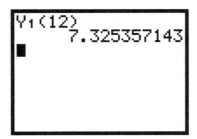

Turn off the STAT PLOT as described on page 9 of this manual before graphing other functions.

THE MAXIMUM AND MINIMUM FEATURES

Section 1.5, Example 2 Use a graphing calculator to determine any relative maxima or minima of the function $f(x) = 0.1x^3 - 0.6x^2 - 0.1x + 2$.

First graph $y_1 = 0.1x^3 - 0.6x^2 - 0.1x + 2$ in a window that displays the relative extrema of the function. Trial and error reveals that one good choice is $[-4, 6, -3, 3]$. Observe that a relative maximum occurs near $x = 0$ and a relative minimum occurs near $x = 4$.

To find the relative maximum, first press $\boxed{\text{2nd}}$ $\boxed{\text{CALC}}$ 4 or $\boxed{\text{2nd}}$ $\boxed{\text{CALC}}$ $\boxed{\triangledown}$ $\boxed{\triangledown}$ $\boxed{\triangledown}$ $\boxed{\text{ENTER}}$ to select the Maximum feature from the CALC menu. We are prompted to select a left bound for the relative maximum. This means that we must choose an x-value that is to the left of the x-value of the point where the relative maximum occurs. This can be done by using the left- and right-arrow keys to move the cursor to a point to the left of the relative maximum or by keying in an appropriate value.

Once this is done, press $\boxed{\text{ENTER}}$. Now we are prompted to select a right bound. We move the cursor to a point to the right of the relative maximum or we can key in an appropriate value.

Press $\boxed{\text{ENTER}}$ again. Finally we are prompted to guess the x-value at which the relative maximum occurs. Move the cursor close to the relative maximum point or key in an x-value.

Press ENTER a third time. We see that a relative maximum function value of approximately 2.004 occurs when $x \approx -0.082$.

To find the relative minimum, select the Minimum feature from the CALC menu by pressing 2nd CALC 3 or 2nd CALC ▽ ▽ ENTER . Select left and right bounds for the relative minimum and guess the x-value at which it occurs as described above. We see that a relative minimum function value of approximately -1.604 occurs when $x \approx 4.082$.

GRAPHING FUNCTIONS DEFINED PIECEWISE

Operations from the TEST menu are used to enter functions that are defined piecewise. (TEST is the second operation associated with the MATH key.) Select DOT mode from the MODE menu when graphing such functions. To select DOT mode press MODE and use ▽ and ▷ to position the blinking cursor over "Dot." Then press ENTER .

We can also select DOT mode by selecting the "dot" GraphStyle on the equation-editor screen. After entering an equation, position the cursor over the GraphStyle icon to the left of the equation and press ENTER repeatedly until the dotted icon appears. If the "line" icon was previously selected, ENTER must be pressed six times to select the "dot" style. Any style selected on the equation-editor screen overrides a MODE selection.

Section 1.5, Example 5 Graph

$$f(x) = \begin{cases} 4, & \text{for } x \le 0, \\ 4 - x^2, & \text{for } 0 < x \le 2, \\ 2x - 6, & \text{for } x > 2. \end{cases}$$

Press $\boxed{Y =}$ and clear any functions that have previously been entered. With the cursor beside "Y1 =" enter the function as described on page 127 of the text by pressing $\boxed{(}$ $\boxed{4}$ $\boxed{)}$ $\boxed{(}$ $\boxed{X, T, \Theta, n}$ $\boxed{2nd}$ $\boxed{TEST}$ $\boxed{6}$ 0 $\boxed{)}$ $\boxed{+}$ $\boxed{(}$ $\boxed{4}$ $\boxed{-}$ $\boxed{X, T, \Theta, n}$ $\boxed{x^2}$ $\boxed{)}$ $\boxed{(}$ $\boxed{0}$ $\boxed{2nd}$ $\boxed{TEST}$ $\boxed{5}$ $\boxed{X, T, \Theta, n}$ $\boxed{)}$ $\boxed{(}$ $\boxed{X, T, \Theta, n}$ $\boxed{2nd}$ $\boxed{TEST}$ $\boxed{6}$ 2 $\boxed{)}$ $\boxed{+}$ $\boxed{(}$ $\boxed{2}$ $\boxed{X, T, \Theta, n}$ $\boxed{-}$ $\boxed{6}$ $\boxed{)}$ $\boxed{(}$ $\boxed{X, T, \Theta, n}$ $\boxed{2nd}$ $\boxed{TEST}$ $\boxed{3}$ 2 $\boxed{)}$. Select a window and then press $\boxed{GRAPH}$. The window $[-5, 5, -3, 6]$ is shown here.

THE GREATEST INTEGER FUNCTION

The greatest integer function is found in the MATH NUM menu and is denoted "int." To find int(1.9) press $\boxed{MATH}$ $\boxed{\triangleright}$ 5 1 $\boxed{.}$ 9 $\boxed{)}$ $\boxed{ENTER}$. Note that the calculator supplies a left parenthesis and we close the parentheses after entering 1.9.

We can also graph the greatest integer function.

Section 1.5, Example 7 Graph $f(x) = \text{int}(x)$.

First we set the calculator in Dot mode as described on page 19 of this manual. If the calculator is left in Connected mode, it will connect adjacent points, or pixels, with line segments and produce a graph of the greatest integer function that looks like stair-steps rather than a set of horizontal line segments.

With the calculator set in DOT mode, press $\boxed{Y =}$ and clear any previously entered functions. Position the cursor beside "Y1 =" and select the greatest integer function from the MATH NUM menu by pressing $\boxed{MATH}$ $\boxed{\triangleright}$ 5 $\boxed{X, T, \Theta, n}$ $\boxed{)}$. Select a window and press $\boxed{GRAPH}$. The window $[-6, 6, -6, 6]$ is shown here.

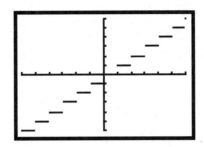

THE COMPOSITION OF FUNCTIONS

We can evaluate composite functions on a graphing calculator.

Section 1.6, Example 4 (b) Given that $f(x) = 2x - 5$ and $g(x) = x^2 - 3x + 8$, find $(f \circ g)(7)$ and $(g \circ f)(7)$.

On the equation-editor screen enter $y_1 = 2x - 5$ and $y_2 = x^2 - 3x + 8$. Then $(f \circ g)(7) = (y_1 \circ y_2)(7)$, or $y_1(y_2(7))$ and $(g \circ f)(7) = (y_2 \circ y_1)(7)$, or $y_2(y_1(7))$. To find these function values press $\boxed{\text{2nd}}$ $\boxed{\text{QUIT}}$ to go to the home screen. Then enter $y_1(y_2(7))$ by pressing $\boxed{\text{VARS}}$ $\boxed{\triangleright}$ 1 1 $\boxed{(}$ $\boxed{\text{VARS}}$ $\boxed{\triangleright}$ 1 2 $\boxed{(}$ 7 $\boxed{)}$ $\boxed{)}$ $\boxed{\text{ENTER}}$. Now enter $y_2(y_1(7))$ by pressing $\boxed{\text{VARS}}$ $\boxed{\triangleright}$ 1 2 $\boxed{(}$ $\boxed{\text{VARS}}$ $\boxed{\triangleright}$ 1 1 $\boxed{(}$ 7 $\boxed{)}$ $\boxed{)}$ $\boxed{\text{ENTER}}$.

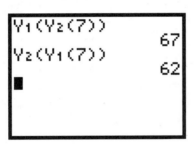

Chapter 2
Functions, Equations, and Inequalities

THE INTERSECT METHOD

We can use the Intersect feature from the CALC menu to solve equations. We call this the **Intersect method**.

Section 2.1, Example 1 Solve $6x - 4 = 1$.

Press $\boxed{Y =}$ to go to the equation-editor screen. Clear any existing entries and then enter $y_1 = 6x - 4$ and $y_2 = 1$. The solution of the original equation is the first coordinate of the point of intersection of the graphs of y_1 and y_2. Graph the equations in a window that displays the point of intersection. The standard window is a good choice.

Next we press $\boxed{\text{2nd}}$ $\boxed{\text{CALC}}$ 5 to select the Intersect feature from the CALC menu. (CALC is the second operation associated with the $\boxed{\text{TRACE}}$ key.) The query "First curve?" appears at the bottom of the screen. The blinking cursor is positioned on the graph of y_1. This is indicated by the notation $Y_1 = 6X - 4$ in the upper left-hand corner of the screen. Press $\boxed{\text{ENTER}}$ to indicate that this is the first curve involved in the intersection. Next the query "Second curve?" appears at the bottom of the screen. The blinking cursor is now positioned on the graph of y_2 and the notation $Y_2 = 1$ should appear in the top left-hand corner of the screen. Press $\boxed{\text{ENTER}}$ to indicate that this is the second curve. We identify the curves for the calculator since we could have as many as ten graphs on the screen at once. After we identify the second curve, the query "Guess?" appears at the bottom of the screen. Use the right and left arrow keys to move the blinking cursor close to the point of intersection of the graphs. This provides the calculator with a guess as to the coordinates of this point. We do this since some pairs of curves can have more than one point of intersection. When the cursor is positioned, press $\boxed{\text{ENTER}}$ a third time. Now the coordinates of the point of intersection appear at the bottom of the screen.

We see that the first coordinate of the point of intersection is 0.83333333. This is a decimal approximation for the coordinate. If it is a rational number, its exact value can be found using the $\triangleright$Frac feature from the MATH menu. To do this first press $\boxed{\text{2nd}}$ $\boxed{\text{QUIT}}$ to go to the home screen. The x- and y-coordinates of the point of intersection are stored in the calculator as X and Y, respectively. To convert the decimal approximation for X to a rational number, press $\boxed{X, T, \Theta, n}$ $\boxed{\text{MATH}}$ 1 $\boxed{\text{ENTER}}$. These keystrokes tell the calculator to use X, and then they access the MATH submenu of the MATH menu, copy item 1 "▷ Frac" to the home screen, and display the conversion. We see that the first coordinate of the point of intersection is $\dfrac{5}{6}$. This is the solution of the equation.

THE ZERO METHOD

The Zero feature of the TI-83, TI-83 Plus, or TI-84 Plus can be used to find the zeros of a function or to solve an equation in the form $f(x) = 0$. We call this the **Zero method**.

Section 2.1, Example 8 Find the zero of $f(x) = 5x - 9$.

On the equation-editor screen, clear any existing entries and then enter $y_1 = 5x - 9$. Now graph the function in a viewing window that shows the x-intercept clearly. The standard window is a good choice.

Press $\boxed{\text{2nd}}$ $\boxed{\text{CALC}}$ to display the CALC menu. Then press 2 to select the Zero feature. We are first prompted to select a left bound for the zero. This means that we must choose an x-value that is to the left of the x-intercept. This can be done by using the left- and right-arrow keys to move the cursor to a point on the curve to the left of the x-intercept or by keying in a value less than the first coordinate of the intercept.

Once this is done press $\boxed{\text{ENTER}}$. Now we are prompted to select a right bound that is to the right of the x-intercept. Again, this can be done by using the arrow keys to move the cursor to a point on the curve to the right of the x-intercept or by keying in a value greater than the first coordinate of the intercept.

Press ENTER again. Finally we are prompted to make a guess as to the value of the zero. Move the cursor to a point close to the zero or key in a value.

Press ENTER a third time. We see that $y = 0$ when $x = 1.8$, so 1.8 is the zero of the function.

If a function has more than one zero, the Zero feature can be used as many times as necessary to find all of them.

OPERATIONS WITH COMPLEX NUMBERS

Operations with complex numbers can be performed on the TI-83, the TI-83 Plus, and the TI-84 Plus. First set the calculator in the complex $a + bi$ mode by pressing MODE, positioning the cursor over $a + bi$, and pressing ENTER.

Section 2.2, Example 2

(a) Add: $(8 + 6i) + (3 + 2i)$.

To find this sum go to the home screen and press 8 [+] 6 [2nd] [i] [+] 3 [+] 2 [2nd] [i] [ENTER]. (The number i is the second operation associated with the [·] key.) Note that it is not necessary to include parentheses when we are adding.

(b) Subtract: $(4 + 5i) - (6 - 3i)$.

Press 4 [+] 5 [2nd] [i] [−] [(] 6 [−] 3 [2nd] [i] [)] [ENTER]. Note that the parentheses must be included as shown so that the entire number $6 - 3i$ is subtracted.

```
8+6i+3+2i
                11+8i
4+5i-(6-3i)
                -2+8i
■
```

Section 2.2, Example 3

(a) Multiply: $\sqrt{-16} \cdot \sqrt{-25}$.

Press [2nd] [$\sqrt{}$] [(−)] 1 6 [)] [2nd] [$\sqrt{}$] [(−)] 2 5 [)] [ENTER].

(b) Multiply: $(1 + 2i)(1 + 3i)$.

Press [(] 1 [+] 2 [2nd] [i] [)] [(] 1 [+] 3 [2nd] [i] [)] [ENTER].

(c) Multiply: $(3 - 7i)^2$.

Press [(] 3 [−] 7 [2nd] [i] [)] [x^2] [ENTER].

```
√(-16)√(-25)
                -20
(1+2i)(1+3i)
              -5+5i
(3-7i)²
            -40-42i
■
```

CHECKING SOLUTIONS OF INEQUALITIES

We can perform a partial check of the solution of an inequality using operations from the TEST menu.

Section 2.6, Example 2 Solve: $-3 < 2x + 5 \leq 7$.

The solution set is found in the text. It is $\{x | -4 < x \leq 1\}$, or $(-4, 1]$. We can perform a partial check of this solution by graphing $y = (-3 < 2x + 5)$ *and* $(2x + 5 \leq 7)$ in Dot mode. (See page 19 in this manual for a discussion of Dot mode.) The value of y will be 1 for those x-values which make y a true statement. It will be 0 for those x-values for which y is false. To enter the expression for y, position the cursor beside Y_1 on the $Y =$ screen. Then press $\boxed{(}$ $\boxed{(-)}$ 3 $\boxed{\text{2nd}}$ $\boxed{\text{TEST}}$ 5 2 $\boxed{\text{X, T, } \Theta, n}$ $\boxed{+}$ 5 $\boxed{)}$ $\boxed{\text{2nd}}$ $\boxed{\text{TEST}}$ $\boxed{\triangleright}$ 1 $\boxed{(}$ 2 $\boxed{\text{X, T, } \Theta, n}$ $\boxed{+}$ 5 $\boxed{\text{2nd}}$ $\boxed{\text{TEST}}$ 6 7 $\boxed{)}$. (TEST is the second operation associated with the $\boxed{\text{MATH}}$ key.) The keystrokes $\boxed{\text{2nd}}$ $\boxed{\text{TEST}}$ 5 display the TEST submenu of the TEST menu and paste the symbol "<" from that menu to the equation-editor screen. The keystrokes $\boxed{\text{2nd}}$ $\boxed{\text{TEST}}$ $\boxed{\triangleright}$ 1 display the LOGIC submenu of the TEST menu and paste "and" from that menu to the equation-editor screen.

Now select a window and press $\boxed{\text{GRAPH}}$. We use the window $[-10, 10, -1, 2]$.

 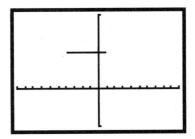

We see that $y = 1$ for x-values from -4 to 1, confirming that all x-values from -4 to 1 are in the solution set. The algebraic solution indicates that the endpoint 1 is also in the solution set.

Chapter 3
Polynomial and Rational Functions

POLYNOMIAL MODELS

We can fit second-degree, or quadratic, functions, third-degree, or cubic, functions and fourth-degree, or quartic, functions to data on a graphing calculator.

Section 3.1, Example 9(a) The table below shows the number of farms, in millions, for several years beginning with 1900. Model the data with both cubic and quartic functions. Let the first coordinate of each data point be the number of years after 1900. Then using R^2, the coefficient of determination, decide which function is the better fit. The closer R^2 is to 1, the better the fit.

Years after 1900	Number of Farms, in millions
0, 1900	5.740
10, 1910	6.366
20, 1920	6.454
30, 1930	6.295
40, 1940	6.102
50, 1950	5.388
60, 1960	3.962
70, 1970	2.954
80, 1980	2.440
90, 1990	2.146
95, 1995	2.196
100, 2000	2.172
102, 2002	2.158

Enter the data in lists as described on page 14 of this manual. To model the data with a cubic function select cubic regression from the STAT CALC menu by pressing $\boxed{\text{STAT}}$ $\boxed{\triangleright}$ $\boxed{6}$ $\boxed{\text{ENTER}}$. The graphing calculator displays the coefficients of a cubic function $y = ax^3 + bx^2 + cx + d$ along with the value of R^2.

After the data are entered, we can also find a quartic model by pressing $\boxed{\text{STAT}}$ $\boxed{\triangleright}$ $\boxed{7}$ $\boxed{\text{ENTER}}$ to select quartic regression from the STAT CALC menu. The calculator displays the coefficients of a quartic function $y = ax^4 + bx^3 + cx^2 + dx + e$

along with the value for R^2.

```
QuarticReg
 y=ax⁴+bx³+…+e
 a=1.4403278E⁻7
 b=⁻7.340139E⁻6
 c=⁻.0018779144
 d=.0822189806
↓e=5.714950699
■
```
```
QuarticReg
 y=ax⁴+bx³+…+e
↑b=⁻7.340139E⁻6
 c=⁻.0018779144
 d=.0822189806
 e=5.714950699
 R²=.9923821873
■
```

Since the R^2-value for the quartic function is closer to 1 than that for the cubic function, the quartic function is the better fit.

A scatterplot of the data can be graphed as described on page 14 of this manual. This function can be copied to the Y = screen using one of the methods described on page 16 of this manual. Then it can be graphed along with the scatterplot. It can also be evaluated using one of the methods on pages 12 and 13.

When appropriate, a quadratic model for data can be found by selecting quadratic regression, item 5, from the STAT CALC menu after the data are entered. This will be demonstrated in Chapter 5.

GRAPHING RATIONAL FUNCTIONS

Section 3.5, Example 1 Consider $f(x) = \dfrac{1}{x-3}$. Find the domain and graph f.

In the text the domain is found to be $\{x \mid x \neq 3\}$, or $(-\infty, 3) \cup (3, \infty)$. Thus, there is not a point on the graph with an x-coordinate of 3. Graphing the function in Connected mode can lead to an incorrect graph in which a line connects the last point plotted to the left of $x = 3$ with the first point plotted to the right of $x = 3$. This line can be eliminated by using Dot mode as described on page 19 of this manual. Selecting a ZDecimal window from the ZOOM menu will also produce a graph in which this line does not appear in Connected mode. To do this, first enter $y = \dfrac{1}{x-3}$ on the Y = screen and then press $\boxed{\text{ZOOM}}$ 4. The resulting window dimensions and graph are shown below.

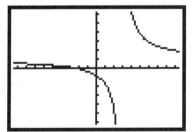

```
WINDOW
 Xmin=⁻4.7
 Xmax=4.7
 Xscl=1
 Ymin=⁻3.1
 Ymax=3.1
 Yscl=1
 Xres=1■
```

Section 3.5, Example 10 Graph: $g(x) = \dfrac{x-2}{x^2-x-2}$.

As explained in the text, the graph of $g(x)$ is the graph of $y = \dfrac{1}{x+1}$ with the point $\left(2, \dfrac{1}{3}\right)$ missing. To use a graphing calculator to produce the graph with a "hole" at $\left(2, \dfrac{1}{3}\right)$, use the ZDecimal feature from the ZOOM menu. After entering $y = \dfrac{x-2}{x^2-x-2}$ on the Y = screen, press $\boxed{\text{ZOOM}}$ 4 to select this window and display the graph.

Chapter 4
Exponential and Logarithmic Functions

GRAPHING AN INVERSE FUNCTION

The DrawInv operation can be used to graph a function and its inverse on the same screen. A formula for the inverse function need not be found in order to do this. The calculator must be set in Func mode when this operation is used.

Section 4.1, Example 7 Graph $f(x) = 2x - 3$ and $f^{-1}(x)$ using the same set of axes.

Enter $y_1 = 2x - 3$ and clear all other functions on the "Y =" screen. Then press $\boxed{\text{2nd}}$ $\boxed{\text{DRAW}}$ 8 to select the DrawInv operation. (DRAW is the second operation associated with the $\boxed{\text{PRGM}}$ key.) Follow these keystrokes with $\boxed{\text{VARS}}$ $\boxed{\triangleright}$ 1 1 to select function y_1. Press $\boxed{\text{ENTER}}$ to see the graph of the function and its inverse. The graphs are shown here in the standard window.

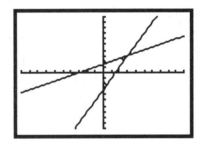

EVALUATING e^x, Log x, and Ln x

Use the calculator's scientific keys to evaluate e^x, $\log x$, and $\ln x$ for specific values of x.

Section 4.2, Example 5 (a), (b) Find the value of e^3 and $e^{-0.23}$. Round to four decimal places.

To find e^3 press $\boxed{\text{2nd}}$ $\boxed{e^x}$ 3 $\boxed{)}$ $\boxed{\text{ENTER}}$. (e^x is the second operation associated with the $\boxed{\text{LN}}$ key.) The calculator returns 20.08553692. Thus, $e^3 \approx 20.0855$. To find $e^{-0.23}$ press $\boxed{\text{2nd}}$ $\boxed{e^x}$ $\boxed{(-)}$ $\boxed{\cdot}$ 2 3 $\boxed{)}$ $\boxed{\text{ENTER}}$. The calculator returns .7945336025, so $e^{-0.23} \approx 0.7945$.

Section 4.3, Example 5 Find the values of log 645,778, log 0.0000239, and log (-3). Round to four decimal places.

To find log 645,778 press $\boxed{\text{LOG}}$ 6 4 5 7 7 8 $\boxed{)}$ $\boxed{\text{ENTER}}$ and read 5.810083246. Thus, log $645,778 \approx 5.8101$. To find log 0.0000239 press $\boxed{\text{LOG}}$ $\boxed{\cdot}$ 0 0 0 0 2 3 9 $\boxed{)}$ $\boxed{\text{ENTER}}$. The calculator returns -4.621602099, so $\log 0.0000239 \approx -4.6216$. When the calculator is set in Real mode the keystrokes $\boxed{\text{LOG}}$ $\boxed{(-)}$ 3 $\boxed{)}$ $\boxed{\text{ENTER}}$ produce the message ERR: NONREAL ANS indicating that the result of this calculation is not a real number.

Section 4.3, Example 6 (a), (b), (c) Find the values of ln 645,778, ln 0.0000239, and ln (−5). Round to four decimal places.

To find ln 645,778 and ln 0.0000239 repeat the keystrokes used above to find log 645,778 and log 0.0000239 but press $\boxed{\text{LN}}$ rather than $\boxed{\text{LOG}}$. We find that ln 645,778 ≈ 13.3782 and ln 0.0000239 ≈ −10.6416. When the calculator is set in Real mode the keystrokes $\boxed{\text{LN}}\ \boxed{(-)}\ 5\ \boxed{)}\ \boxed{\text{ENTER}}$ produce the message ERR: NONREAL ANS indicating that the result of this calculation is not a real number.

USING THE CHANGE OF BASE FORMULA

To find a logarithm with a base other than 10 or e we use the change-of-base formula, $\log_b M = \dfrac{\log_a M}{\log_a b}$, where a and b are any logarithmic bases and M is any positive number.

Section 4.3, Example 7 Find $\log_5 8$ using common logarithms.

We let $a = 10$, $b = 5$, and $M = 8$ and substitute in the change-of-base formula. Press $\boxed{\text{LOG}}\ 8\ \boxed{)}\ \boxed{\div}\ \boxed{\text{LOG}}\ 5\ \boxed{)}$ $\boxed{\text{ENTER}}$. Note that the parentheses must be closed in the numerator to enter the expression correctly. We also close the parentheses in the denominator for completeness. The result is about 1.2920. We could have let $a = e$ and used natural logarithms to find $\log_5 8$ as well.

```
log(8)/log(5)
          1.292029674
■
```

Section 4.3, Example 9 Graph $y = \log_5 x$.

To use a graphing calculator we must first change the base to e or 10. Here we use e. Let $a = e$, $b = 5$, and $M = x$ and substitute in the change-of-base formula. Enter $y_1 = \dfrac{\ln x}{\ln 5}$ on the Y = screen, select a window, and press $\boxed{\text{GRAPH}}$. Note that since the TI-83, the TI-83 Plus, and the TI-84 Plus force the use of parentheses with the ln function, the parentheses in the numerator must be closed on these calculators: $\ln(x)/\ln(5)$. The right parenthesis following the 5 is optional but we include it for completeness.

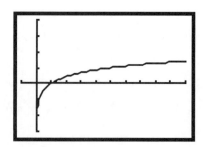

EXPONENTIAL AND LOGARITHMIC REGRESSION

In addition to the types of polynomial regression discussed earlier, exponential and logarithmic functions can be fit to data. The operations of entering data, making scatterplots, and graphing and evaluating these functions are the same as for linear regression functions. So are the procedures for copying a regression equation to the Y = screen, graphing it, and using it to find function values. Note that the coefficient of correlation, r, will be displayed only if DiagnosticOn has been selected from the Catalog.

Section 4.6, Example 6 (a) *Debit Card Volume.* The total debit card volume has increased dramatically in recent years, as shown in the table below. Use a graphing calculator to fit an exponential function to the data.

| | Debit Card Volume, y |
Year, x	(in billions)
1997, 0	162.5
1998, 1	238.5
1999, 2	322.6
2000, 3	419.1
2001, 4	571.8

Enter the data in lists as described on page 14 of this manual. Then select exponential regression from the STAT CALC menu by pressing STAT ▷ 0 ENTER. The calculator displays the coefficient a and the base b for the exponential function $y = a \cdot b^x$.

A scatterplot of the data can be graphed as described on page 14 of this manual. This function can be copied to the Y = screen using one of the methods described on page 16 of this manual. Then it can be graphed along with the scatterplot. It can also be evaluated using one of the methods on pages 12 and 13.

Section 4.6, Exercise 28 (a) *Forgetting.* In an art class, students were tested at the end of the course on a final exam. Then they were retested with an equivalent test at subsequent time intervals. Their scores after time x, in months, are given in the following table. Use a graphing calculator to fit a logarithmic function $y = a + b \ln x$ to the data.

Time, x (in months)	Score, y
1	84.9%
2	84.6%
3	84.4%
4	84.2%
5	84.1%
6	83.9%

After entering the data in lists as described on page 14 of this manual, press $\boxed{\text{STAT}}$ $\boxed{\triangleright}$ to view the STAT CALC menu. Select LnReg by pressing 9 $\boxed{\text{ENTER}}$. The values of a and b for the logarithmic function $y = a + b \ln x$ are displayed.

LOGISTIC REGRESSION

A logistic function can be fit to data using the TI-83, Ti-83 Plus, or TI-84 Plus.

Section 4.6, Exercise 32 (a) *Effect of Advertising.* A company introduces a new software product on a trial run in a city. They advertised the product on television and found the following data relating the percent P of people who bought the product after x ads were run. Use a graphing calculator to fit a logistic function $P(x) = \dfrac{a}{1 + be^{-kx}}$ to the data.

Number of Ads, x	Percent Who Bought, P
0	0.2
10	0.7
20	2.7
30	9.2
40	27
50	57.6
60	83.3
70	94.8
80	98.5
90	99.6

After entering the data in lists as described on page 14 of this manual, press $\boxed{\text{STAT}}$ $\boxed{\triangleright}$ to view the STAT CALC menu. Select Logistic by pressing $\boxed{\text{ALPHA}}$ B $\boxed{\text{ENTER}}$. The values of a, b, and c for the logistic function $y = \dfrac{c}{1 + ae^{-bx}}$ are displayed.

```
Logistic
 y=c/(1+ae^(-bx))
 a=489.2438401
 b=.1299899024
 c=99.98884912
```

Chapter 5
Systems of Equations and Matrices

QUADRATIC REGRESSION

Quadratic functions can be fit to data using the quadratic regression operation from the STAT CALC menu. The operations of entering data, making scatterplots, and graphing and evaluating quadratic regression functions are the same as for the other regression functions discussed earlier.

Section 5.2, Exercise 35 *Morning Newspapers* The number of morning newspapers in the United States in various years is shown in the following table.

Year	Number of Morning Newspapers
1920	437
1940	380
1960	312
1980	387
1990	559
2001	776

(a) Use a graphing calculator to fit a quadratic function to the data, where x is the number of years after 1920.

(b) Use the function found in part (a) to estimate the number of morning newspapers in 2006 and 2009.

(a) Clear any existing entries on the equation-editor screen. Then enter the data in L_1 and L_2 as described on page 14 of this manual.

To fit a quadratic function to the data, press $\boxed{\text{STAT}}$ $\boxed{\triangleright}$ to view the STAT CALC menu. Then select QuadReg by pressing 5 $\boxed{\text{ENTER}}$. The coefficients of a quadratic equation $y = ax^2 + bx + c$ are displayed. Note that at least three data points are required for quadratic regression.

(b) To estimate the number of morning newspapers in 2006 and 2009, we evaluate the regression function for $x = 86$ and $x = 89$, respectively. We can use any of the methods for evaluating a function found on pages 12 and 13 of this manual.

Here we show a table set in Ask mode. We estimate that there will be about 836 morning newspapers in 2006 and about 897 in 2009.

MATRICES AND ROW-EQUIVALENT OPERATIONS

Matrices with up to 99 rows or columns can be entered on a graphing calculator. As many as ten matrices can be entered on the TI-83, Ti-83 Plus, or TI-84 Plus at one time. Row-equivalent operations can be performed on matrices on the calculator.

Section 5.3, Example 1 Solve the following system:

$$\begin{aligned} 2x - y + 4z &= -3, \\ x - 2y - 10z &= -6, \\ 3x \quad + 4z &= 7. \end{aligned}$$

First we enter the augmented matrix

$$\begin{bmatrix} 2 & -1 & 4 & -3 \\ 1 & -2 & -10 & -6 \\ 3 & 0 & 4 & 7 \end{bmatrix}$$

in the graphing calculator. Begin by pressing $\boxed{\text{2nd}}$ $\boxed{\text{MATRX}}$ $\boxed{\triangleright}$ $\boxed{\triangleright}$ to display the MATRIX EDIT menu on the TI-83 Plus. This key is labeled $\boxed{\text{MATRIX}}$ on the TI-84 Plus. We will use MATRX and MATRIX interchageably in this manual. (MATRX/MATRIX is the second operation associated with the $\boxed{x^{-1}}$ key.) On the TI-83 simply press $\boxed{\text{MATRX}}$ $\boxed{\triangleright}$ $\boxed{\triangleright}$. Then select the matrix to be defined. We will select matrix $[\mathbf{A}]$ by pressing 1. Now the MATRIX EDIT screen appears. The dimensions of the matrix are displayed on the top line of this screen, with the cursor on the row dimension. Enter the dimensions of the augmented matrix, 3 x 4, by pressing 3 $\boxed{\text{ENTER}}$ 4 $\boxed{\text{ENTER}}$. Now the cursor moves to the element in the first row and first column of the matrix. Enter the elements of the first row by pressing 2 $\boxed{\text{ENTER}}$ $\boxed{(-)}$ 1 $\boxed{\text{ENTER}}$ 4 $\boxed{\text{ENTER}}$ $\boxed{(-)}$ 3 $\boxed{\text{ENTER}}$. The cursor moves to the element in the second row and first column of the matrix. Enter the elements of the second and third rows of the augmented matrix by typing each in turn followed by $\boxed{\text{ENTER}}$ as above. Note that the screen only displays three columns of the matrix. The arrow keys can be used to move the cursor to any element at any time.

Row-equivalent operations are performed by making selections from the MATRIX MATH menu. To view this menu press $\boxed{\text{2nd}}$ $\boxed{\text{QUIT}}$ to leave the MATRIX EDIT screen. Then press $\boxed{\text{2nd}}$ $\boxed{\text{MATRX}}$ $\boxed{\triangleright}$ on the TI-83 Plus or TI-84 Plus or $\boxed{\text{MATRX}}$ $\boxed{\triangleright}$ on the TI-83. Now press $\boxed{\triangledown}$ fifteen times or $\boxed{\triangle}$ one time to see the four row-equivalent operations, C: rowSwap(, D: row+(, E: *row(, and F: *row+(. These operations interchange two rows of a matrix, add two rows, multiply a row by a number, and multiply a row by a number and add it to a second row, respectively.

We will use the calculator to perform the row-equivalent operations that were done algebraically in the text. First, to interchange row 1 and row 2 of matrix [**A**], with the MATRIX MATH menu displayed, press $\boxed{\text{ALPHA}}$ $\boxed{\text{C}}$ to select rowSwap. Then press $\boxed{\text{2nd}}$ $\boxed{\text{MATRX}}$ 1 on the TI-83 Plus or TI-84 Plus or $\boxed{\text{MATRX}}$ 1 on the TI-83 to select [**A**]. Follow this with a comma and the rows to be interchanged, $\boxed{\text{,}}$ 1 $\boxed{\text{,}}$ 2 $\boxed{\text{)}}$ $\boxed{\text{ENTER}}$.

```
rowSwap([A],1,2)
[[1  -2  -10  -6]
 [2  -1   4   -3]
 [3   0   4    7 ]]
```

The graphing calculator will not store the matrix produced using a row-equivalent operation, so when several operations are to be performed in succession it is helpful to store the result of each operation as it is produced. For example, to store the matrix resulting from interchanging the first and second rows of [**A**] as matrix [**B**] on the TI-83 Plus or TI-84 Plus press $\boxed{\text{STO}\triangleright}$ $\boxed{\text{2nd}}$ $\boxed{\text{MATRX}}$ 2 $\boxed{\text{ENTER}}$ immediately after interchanging the rows. Press $\boxed{\text{STO}\triangleright}$ $\boxed{\text{MATRX}}$ 2 $\boxed{\text{ENTER}}$ on the TI-83. This can also be done before $\boxed{\text{ENTER}}$ is pressed at the end of the rowSwap.

Next we multiply the first row of [**B**] by -2, add it to the second row and store the result as [**B**] again on the TI-83 Plus or TI-84 Plus by pressing $\boxed{\text{2nd}}$ $\boxed{\text{MATRX}}$ $\boxed{\triangleright}$ $\boxed{\text{ALPHA}}$ $\boxed{\text{F}}$ $\boxed{(-)}$ 2 $\boxed{\text{,}}$ $\boxed{\text{2nd}}$ $\boxed{\text{MATRX}}$ 2 $\boxed{\text{,}}$ 1 $\boxed{\text{,}}$ 2 $\boxed{\text{)}}$ $\boxed{\text{STO}\triangleright}$ $\boxed{\text{2nd}}$ $\boxed{\text{MATRX}}$ 2 $\boxed{\text{ENTER}}$. It is not necessary to press $\boxed{\text{2nd}}$ before $\boxed{\text{MATRX}}$ on the TI-83. These keystrokes select *row+(from the MATRIX MATH menu; then they specify that the value of the multiplier is -2, the matrix being operated on is [**B**], and that a multiple of row 1 is being added to row 2; finally they store the result as [**B**].

To multiply row 1 by -3, add it to row 3, and store the result as [**B**] on the TI-83 Plus or TI-84 Plus press $\boxed{\text{2nd}}$ $\boxed{\text{MATRX}}$ $\boxed{\triangleright}$ $\boxed{\text{ALPHA}}$ $\boxed{\text{F}}$ $\boxed{(-)}$ 3 $\boxed{\text{,}}$ $\boxed{\text{2nd}}$ $\boxed{\text{MATRX}}$ 2 $\boxed{\text{,}}$ 1 $\boxed{\text{,}}$ 3 $\boxed{\text{)}}$ $\boxed{\text{STO}\triangleright}$ $\boxed{\text{2nd}}$ $\boxed{\text{MATRX}}$ 2 $\boxed{\text{ENTER}}$. It is not necessary to press $\boxed{\text{2nd}}$ before $\boxed{\text{MATRX}}$ on the TI-83.

```
*row+(-3,[B],1,3
)→[B]
[[1  -2  -10  -6]
 [0   3   24   9 ]
 [0   6   34   25]]
```

Now multiply the second row by 1/3 and store the result as [**B**] again. On the TI-83 Plus or TI-84 Plus press $\boxed{\text{2nd}}$

MATRX $\triangleright$ ALPHA E 1 $(\div)$ 3 $,$ 2nd MATRX 2 $,$ 2 $)$ STO$\triangleright$ 2nd MATRX 2 ENTER . It is not necessary to press 2nd before MATRX on the TI-83. These keystrokes select *row(from the MATRIX MATH menu; then they specify that the value of the multiplier is 1/3, the matrix being operated on is [**B**], and row 2 is being multiplied; finally they store the result as [**B**]. The keystrokes 1 $(\div)$ 3 could be replaced with 3 x^{-1} .

```
*row(1/3,[B],2)→
[B]
 [[1  -2 -10 -6]
  [0  1   8   3 ]
  [0  6   34  25]]
```

Multiply the second row by -6 and add it to the third row using *row+(. On the TI-83 Plus or TI-84 Plus press 2nd MATRX $\triangleright$ ALPHA F $(-)$ 6 $,$ 2nd MATRX 2 $,$ 2 $,$ 3 $)$ STO$\triangleright$ 2nd MATRX 2 ENTER . It is not necessary to press 2nd before MATRX on the TI-83. QUIT to leave this screen.

```
*row+(-6,[B],2,3
)→[B]
 [[1  -2 -10 -6]
  [0  1   8   3 ]
  [0  0  -14  7 ]]
```

Finally, multiply the third row by $-1/14$ on the TI-83 Plus or TI-84 Plus by pressing 2nd MATRX $\triangleright$ ALPHA E $(-)$ 1 $(\div)$ 1 4 $,$ 2nd MATRX 2 $,$ 3 $)$ ENTER . It is not necessary to press 2nd before MATRX on the TI-83. The keystrokes $(-)$ 1 $(\div)$ 1 4 could be replaced with $(-)$ 1 4 x^{-1} .

```
*row(-1/14,[B],3
)
 [[1  -2 -10 -6 ]
  [0  1   8   3  ]
  [0  0   1  -.5]]
```

Write the system of equations that corresponds to the final matrix. Then use back-substitution to solve for x, y, and z as illustrated in the text.

Section 5.3, Example 3 Instead of stopping with row-echelon form as we did in Example 1 above, we can continue to apply row-equivalent operations until the matrix is in reduced row-echelon form. Reduced row-echelon form of a matrix can be found directly by using the rref(operation from the MATRIX MATH menu. For example, to find reduced row-echelon form for matrix **A** in Example 1 above, after entering [**A**] and leaving the MATRIX EDIT screen on the TI-83 Plus press $\boxed{\text{2nd}}$ $\boxed{\text{MATRX}}$ $\boxed{\triangleright}$ $\boxed{\text{ALPHA}}$ $\boxed{\text{B}}$ $\boxed{\text{2nd}}$ $\boxed{\text{MATRX}}$ 1 $\boxed{\text{ENTER}}$. It is not necessary to press $\boxed{\text{2nd}}$ before $\boxed{\text{MATRX}}$ on the TI-83. We can read the solution of the system of equations, $(3, 7, -0.5)$ directly from the resulting matrix.

```
rref([A]
     [[1 0 0 3   ]
      [0 1 0 7   ]
      [0 0 1 -.5]]
```

MATRIX OPERATIONS

We can use a graphing calculator to add and subtract matrices, to multiply a matrix by a scalar, and to multiply matrices.

Section 5.4, Example 1 (a) Find **A** + **B** for

$$\mathbf{A} = \begin{bmatrix} -5 & 0 \\ 4 & \frac{1}{2} \end{bmatrix}, \ \mathbf{B} = \begin{bmatrix} 6 & -3 \\ 2 & 3 \end{bmatrix}.$$

Enter **A** and **B** on the MATRIX EDIT screen as [**A**] and [**B**] as described earlier in this chapter of the Graphing Calculator Manual. Press $\boxed{\text{2nd}}$ $\boxed{\text{QUIT}}$ to leave this screen. Then press $\boxed{\text{2nd}}$ $\boxed{\text{MATRX}}$ 1 $\boxed{+}$ $\boxed{\text{2nd}}$ $\boxed{\text{MATRX}}$ 2 $\boxed{\text{ENTER}}$ to display the sum. It is not necessary to press $\boxed{\text{2nd}}$ before $\boxed{\text{MATRX}}$ on the TI-83.

```
[A]+[B]
      [[1 -3  ]
       [6 3.5]]
```

Section 5.4, Example 2 Find **C** − **D** for each of the following.

a) $\mathbf{C} = \begin{bmatrix} 1 & 2 \\ -2 & 0 \\ -3 & -1 \end{bmatrix}, \ \mathbf{D} = \begin{bmatrix} 1 & -1 \\ 1 & 3 \\ 2 & 3 \end{bmatrix}$ 　　b) $\mathbf{C} = \begin{bmatrix} 5 & -6 \\ -3 & 4 \end{bmatrix}, \ \mathbf{D} = \begin{bmatrix} -4 \\ 1 \end{bmatrix}$

a) Enter **C** and **D** on the MATRIX EDIT screen as [**C**] and [**D**]. Press $\boxed{\text{2nd}}$ $\boxed{\text{QUIT}}$ to leave this screen. Then press $\boxed{\text{2nd}}$ $\boxed{\text{MATRX}}$ 3 $\boxed{-}$ $\boxed{\text{2nd}}$ $\boxed{\text{MATRX}}$ 4 $\boxed{\text{ENTER}}$ to display the difference on the TI-83 Plus or TI-84 Plus. It is not necessary to press $\boxed{\text{2nd}}$ before $\boxed{\text{MATRX}}$ on the TI-83.

b) Enter **C** and **D** on the MATRIX EDIT screen as [**C**] and [**D**]. Press 2nd QUIT to leave this screen. Then, on the TI-83 Plus or TI-84 Plus, press 2nd MATRX 3 − 2nd MATRX 4 ENTER. It is not necessary to press 2nd before MATRX on the TI-83. The calculator returns the message ERR:DIM MISMATCH, indicating that this subtraction is not possible. This is the case because the matrices have different orders.

Section 5.4, Example 4 Find $3\mathbf{A}$ and $(-1)\mathbf{A}$, for $\mathbf{A} = \begin{bmatrix} -3 & 0 \\ 4 & 5 \end{bmatrix}$.

Enter **A** on the MATRIX EDIT screen as [**A**]. Press 2nd QUIT to leave this screen. Then to find $3\mathbf{A}$ on the TI-83 Plus or TI-84 Plus press 3 2nd MATRX 1 ENTER and to find $(-1)\mathbf{A}$ press (−) 1 2nd MATRX 1 ENTER. Note that $(-1)\mathbf{A}$ is the opposite, or additive inverse, of **A** and can also be found by pressing (−) 2nd MATRX 1 ENTER. It is not necessary to press 2nd before MATRX on the TI-83.

Section 5.4, Example 6 (a), (d) For

$$\mathbf{A} = \begin{bmatrix} 3 & 1 & -1 \\ 2 & 0 & 3 \end{bmatrix}, \mathbf{B} = \begin{bmatrix} 1 & 6 \\ 3 & -5 \\ -2 & 4 \end{bmatrix}, \text{ and } \mathbf{C} = \begin{bmatrix} 4 & -6 \\ 1 & 2 \end{bmatrix}$$

find each of the following.

a) **AB**   d) **AC**

First enter **A**, **B**, and **C** as [**A**], [**B**], and [**C**] on the MATRIX EDIT screen. Press $\boxed{\text{2nd}}$ $\boxed{\text{QUIT}}$ to leave this screen.

a) To find **AB** on the TI-83 Plus or TI-84 Plus press $\boxed{\text{2nd}}$ $\boxed{\text{MATRX}}$ 1 $\boxed{\text{2nd}}$ $\boxed{\text{MATRX}}$ 2 $\boxed{\text{ENTER}}$. It is not necessary to press $\boxed{\text{2nd}}$ before $\boxed{\text{MATRX}}$ on the TI-83.

```
[A] [B]
         [[8   9 ]
          [-4  24]]
```

d) To find **AC** on the TI-83 Plus or TI-84 Plus press $\boxed{\text{2nd}}$ $\boxed{\text{MATRX}}$ 1 $\boxed{\text{2nd}}$ $\boxed{\text{MATRX}}$ 3 $\boxed{\text{ENTER}}$. It is not necessary to press $\boxed{\text{2nd}}$ before $\boxed{\text{MATRX}}$ on the TI-83. The calculator returns the message ERR:DIM MISMATCH, indicating that this multiplication is not possible. This is the case because the number of columns in **A** is not the same as the number of rows in **C**. Thus, the matrices cannot be multiplied in this order.

FINDING THE INVERSE OF A MATRIX

The inverse of a matrix can be found quickly on a graphing calculator.

Section 5.5, Example 3 Find $\mathbf{A}^{-1}$, where

$$\mathbf{A} = \begin{bmatrix} -2 & 3 \\ -3 & 4 \end{bmatrix}.$$

Enter **A** as [**A**] on the MATRIX EDIT screen. Then press $\boxed{\text{2nd}}$ $\boxed{\text{QUIT}}$ to leave this screen. Now, on the TI-83 Plus or TI-84 Plus, press $\boxed{\text{2nd}}$ $\boxed{\text{MATRX}}$ 1 $\boxed{x^{-1}}$ $\boxed{\text{ENTER}}$. It is not necessary to press $\boxed{\text{2nd}}$ before $\boxed{\text{MATRX}}$ on the TI-83.

```
[A]-1
         [[4  -3]
          [3  -2]]
```

Section 5.5, Exercise 7 Find $\mathbf{A}^{-1}$, where

$$\mathbf{A} = \begin{bmatrix} 6 & 9 \\ 4 & 6 \end{bmatrix}.$$

Enter $\mathbf{A}$ as [A] on the MATRIX EDIT screen and then press $\boxed{\text{2nd}}$ $\boxed{\text{QUIT}}$ to leave this screen. Now, on the TI-83 Plus or TI-84 Plus, press $\boxed{\text{2nd}}$ $\boxed{\text{MATRX}}$ 1 $\boxed{x^{-1}}$ $\boxed{\text{ENTER}}$. It is not necessary to press $\boxed{\text{2nd}}$ before $\boxed{\text{MATRX}}$ on the TI-83. The calculator returns the message ERR:SINGULAR MAT, indicating that $\mathbf{A}^{-1}$ does not exist.

MATRIX SOLUTIONS OF SYSTEMS OF EQUATIONS

We can write a system of n linear equations in n variables as a matrix equation $\mathbf{AX} = \mathbf{B}$. If $\mathbf{A}$ has an inverse the solution of the system of equations is given by $\mathbf{X} = \mathbf{A}^{-1}\mathbf{B}$.

Section 5.5, Example 5 Use an inverse matrix to solve the following system of equations:

$$-2x + 3y = 4,$$
$$-3x + 4y = 5.$$

Enter $\mathbf{A} = \begin{bmatrix} -2 & 3 \\ -3 & 4 \end{bmatrix}$ and $\mathbf{B} = \begin{bmatrix} 4 \\ 5 \end{bmatrix}$ on the MATRIX EDIT screen as [A] and [B]. Press $\boxed{\text{2nd}}$ $\boxed{\text{QUIT}}$ to leave this screen. Then, on the TI-83 Plus or TI-84 Plus, press $\boxed{\text{2nd}}$ $\boxed{\text{MATRX}}$ 1 $\boxed{x^{-1}}$ $\boxed{\text{2nd}}$ $\boxed{\text{MATRX}}$ 2 $\boxed{\text{ENTER}}$. It is not necessary to press $\boxed{\text{2nd}}$ before $\boxed{\text{MATRX}}$ on the TI-83. The result is the 2 x 1 matrix $\begin{bmatrix} 1 \\ 2 \end{bmatrix}$, so the solution is $(1, 2)$.

DETERMINANTS AND CRAMER'S RULE

We can evaluate determinants on a graphing calculator and use Cramer's rule to solve systems of equations.

Section 5.6, Example 5 Use a graphing calculator to evaluate $|\mathbf{A}|$.

$$\mathbf{A} = \begin{bmatrix} 1 & 6 & -1 \\ -3 & -5 & 3 \\ 0 & 4 & 2 \end{bmatrix}$$

Enter $\mathbf{A}$ on the MATRIX EDIT screen and then press $\boxed{\text{2nd}}$ $\boxed{\text{QUIT}}$ to leave this screen. We will select the "det(" operation from the MATRIX MATH menu. On the TI-83 Plus or TI-84 Plus, press $\boxed{\text{2nd}}$ $\boxed{\text{MATRX}}$ $\boxed{\triangleright}$ 1 $\boxed{\text{2nd}}$ $\boxed{\text{MATRX}}$ 1 $\boxed{)}$ $\boxed{\text{ENTER}}$. It is not necessary to press $\boxed{\text{2nd}}$ before pressing $\boxed{\text{MATRX}}$ on the TI-83. We see that $|\mathbf{A}| = 26$.

Section 5.6, Example 6 Solve using Cramer's rule:

$$2x + 5y = 7,$$
$$5x - 2y = -3.$$

First we enter the matrices corresponding to D, D_x, and D_y as **A**, **B**, and **C**, respectively. We have

$$\mathbf{A} = \begin{bmatrix} 2 & 5 \\ 5 & -2 \end{bmatrix}, \mathbf{B} = \begin{bmatrix} 7 & 5 \\ -3 & -2 \end{bmatrix}, \text{ and } \mathbf{C} = \begin{bmatrix} 2 & 7 \\ 5 & -3 \end{bmatrix}.$$

Then we use the "det(" operation from the MATRIX MATH menu. We have

$$x = \frac{\det(\mathbf{B})}{\det(\mathbf{A})} \text{ and } y = \frac{\det(\mathbf{C})}{\det(\mathbf{A})}.$$

To find x expressed in fraction notation on the TI-83 Plus or TI-84 Plus, press 2nd MATRX ▷ 1 2nd MATRX 2) ÷ 2nd MATRX ▷ 1 2nd MATRX 1) MATH 1 ENTER. It is not necessary to press 2nd before pressing MATRX on the TI-83. The result is $-\dfrac{1}{29}$. To find y expressed in fraction notation on the TI-83 Plus or TI-84 Plus, press 2nd MATRX ▷ 1 2nd MATRX 3) ÷ 2nd MATRX ▷ 1 2nd MATRX 1) MATH 1 ENTER. We can also recall the entry for x and edit it, replacing matrix **B** with matrix **C** to find y. In either case the result is $\dfrac{41}{29}$. As before, it is not necessary to press 2nd before pressing MATRX on the TI-83. The solution of the system of equations is $\left(-\dfrac{1}{29}, \dfrac{41}{29}\right)$.

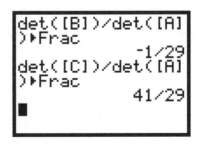

GRAPHS OF INEQUALITIES

We can graph linear inequalities on a graphing calculator, shading the region of the solution set. The calculator should be set in Func mode at this point.

Section 5.7, Example 1 Graph: $y < x + 3$.

First we graph the related equation $y = x + 3$. We use the standard window $[-10, 10, -10, 10]$. Since the inequality symbol is $<$ we know that the line $y = x + 3$ is not part of the solution set. In a hand-drawn graph we would use a dashed line to indicate this. We can select Dot mode or use the Dot GraphStyle. (See page 19 of this manual for instructions on selecting Dot mode and Dot GraphStyle.) Even if Dot mode or graphstyle is used, however, the line appears to be solid. After determining that the solution set of the inequality consists of all points below the line, we use the calculator's SHADE operation to shade this region. SHADE is item 7 on the DRAW DRAW menu. To access it press $\boxed{\text{2nd}}$ $\boxed{\text{DRAW}}$ 7.

Now enter a lower function and an upper function. The region between them will be shaded. We want to shade the area between the bottom of the window, $y = -10$, and the line $y = x + 3$ so we enter $\boxed{(-)}$ 1 0 $\boxed{,}$ $\boxed{\text{X,T,}\Theta,n}$ $\boxed{+}$ 3 $\boxed{)}$ $\boxed{\text{ENTER}}$. We can also enter $x + 3$ as y_1. The result is shown below. Keep in mind that the line $y = x + 3$ is not included in the solution set.

The "shade below" GraphStyle can also be used to shade this region. After entering the related equation, cycle through the Graph Style options on the "Y =" screen as described on page 19 of this manual until the "shade below" option appears. Then press $\boxed{\text{GRAPH}}$ to display the graph of the inequality. As mentioned above, keep in mind the fact that the line $y = x + 3$ is not included in the solution set.

            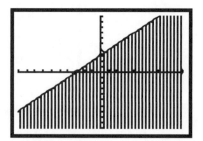

We can also use the SHADE operation from the DRAW menu to graph a system of inequalities when the solution set lies between the graphs of two functions.

Section 5.7, Exercise 43 Graph:
$$y \leq x,$$
$$y \geq 3 - x.$$

First graph the related equations $y_1 = x$ and $y_2 = 3 - x$ and determine that the solution set consists of all the points on or above the graph of $y_2 = 3 - x$ and on or below the graph of $y_1 = x$. We will shade this region by pressing $\boxed{\text{2nd}}$ $\boxed{\text{DRAW}}$ 7 3 $\boxed{-}$ $\boxed{\text{X,T,}\Theta,n}$ $\boxed{,}$ $\boxed{\text{X,T,}\Theta,n}$ $\boxed{)}$ $\boxed{\text{ENTER}}$. These keystrokes select the SHADE operation from the DRAW DRAW menu and then enter $y_2 = 3 - x$ as the lower function and $y_1 = x$ as the upper function. We could also enter these functions as y_2 and y_1, respectively.

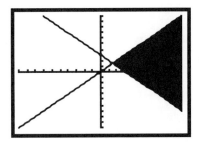

We can also graph systems of inequalities by shading the solution set of each inequality in the system with a different pattern. When the "shade above" or "shade below" Graph Style options are selected the calculator rotates through four shading patterns. Vertical lines shade the first function, horizontal lines the second, negatively sloping diagonal lines the third, and positively sloping diagonal lines the fourth. These patterns repeat if more than four functions are graphed.

Section 5.7, Example 5 Graph the solution set of the system
$$x + y \leq 4,$$
$$x - y \geq 2.$$

First graph the equation $x + y = 4$, entering it in the form $y_1 = -x + 4$. We determine that the solution set of $x + y \leq 4$ consists of all points below the line $x + y = 4$, or $y_1 = -x + 4$, so we select the "shade below" GraphStyle for this function. Next graph $x - y = 2$, entering it in the form $y_2 = x - 2$. The solution set of $x - y \geq 2$ is all points below the line $x - y = 2$, or $y_2 = x - 2$, so we also choose the "shade below" GraphStyle for this function. Now press $\boxed{\text{GRAPH}}$ to display the solution sets of each inequality in the system and the region where they overlap. The region of overlap is the solution set of the system of inequalities.

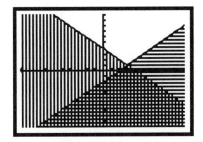

Chapter 6
Conic Sections

Many conic sections are represented by equations that are not functions. Consequently, these equations must be entered on the TI-83, TI-83 Plus, and TI-84 Plus as two equations, each of which is a function.

GRAPHING PARABOLAS

To graph a parabola of the form $y^2 = 4px$ or $(y - k)^2 = 4p(x - h)$, we must first solve the equation for y.

Section 6.1, Example 4 Graph the parabola $y^2 - 2y - 8x - 31 = 0$.

In the text we used the quadratic formula to solve the equation for y:
$$y = \frac{2 \pm \sqrt{32x + 128}}{2}.$$

One way to produce the graph of the parabola is to enter $y_1 = \dfrac{2 + \sqrt{32x + 128}}{2}$ and $y_2 = \dfrac{2 - \sqrt{32x + 128}}{2}$, select a window, and press $\boxed{\text{GRAPH}}$ to see the graph. Here we use the square window $[-12, 12, -8, 8]$. The first equation produces the top half of the parabola and the second equation produces the lower half.

 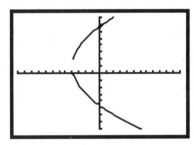

We can also enter $y_1 = \sqrt{32x + 128}$ and then enter $y_2 = \dfrac{2 + y_1}{2}$ and $y_3 = \dfrac{2 - y_1}{2}$. For example, to enter $y_2 = \dfrac{2 + y_1}{2}$ position the cursor beside "Y2 =" and press $\boxed{(}$ $\boxed{2}$ $\boxed{+}$ $\boxed{\text{VARS}}$ $\boxed{\triangleright}$ $\boxed{1}$ $\boxed{1}$ $\boxed{)}$ $\boxed{\div}$ 2. Enter $y_3 = \dfrac{2 - y_1}{2}$ in a similar manner. Finally, deselect y_1 by moving the cursor to the equals sign following Y1 and pressing $\boxed{\text{ENTER}}$. Note that the equals sign in this equation is no longer highlighted. This indicates that it has been deselected and thus its graph will not appear with the graphs of the equations that remain selected.

The top half of the graph is produced by y_2 and the lower half by y_3. The expression for y_1 was entered to avoid entering the radical expression more than once. By deselecting y_1 we prevent its graph from appearing on the screen with the graph of the parabola.

```
Plot1 Plot2 Plot3
\Y1=√(32X+128)
\Y2⊟(2+Y1)/2
\Y3⊟(2-Y1)/2█
\Y4=
\Y5=
\Y6=
\Y7=
```

We could also use the standard equation of the parabola found in the text:

$$(y-1)^2 = 8(x+4).$$

Solve this equation for y.

$$y - 1 = \pm\sqrt{8(x+4)}$$
$$y = 1 \pm \sqrt{8(x+4)}$$

Then enter $y_1 = 1 + \sqrt{8(x+4)}$ and $y_2 = 1 - \sqrt{8(x+4)}$, or enter $y_1 = \sqrt{8(x+4)}$, $y_2 = 1 + y_1$, and $y_3 = 1 - y_1$, and deselect y_1 as described above.

```
Plot1 Plot2 Plot3
\Y1⊟1+√(8(X+4))
\Y2⊟1-√(8(X+4))█
\Y3=
\Y4=
\Y5=
\Y6=
\Y7=
```

```
Plot1 Plot2 Plot3
\Y1=√(8(X+4))
\Y2⊟1+Y1█
\Y3⊟1-Y1█
\Y4=
\Y5=
\Y6=
\Y7=
```

GRAPHING CIRCLES

The equation of a circle must be solved for y before it can be entered on a TI-83, TI-83 Plus, or TI-84 Plus.

Section 6.2, Example 1 Graph the circle $x^2 + y^2 - 16x + 14y + 32 = 0$.

In the text we found the standard form for the equation of the circle and then solved for y:

$$y = -7 \pm \sqrt{81 - (x-8)^2}.$$

We could also have solved the original equation using the quadratic formula.

One way to produce the graph is to enter $y_1 = -7 + \sqrt{81 - (x-8)^2}$ and $y_2 = -7 - \sqrt{81 - (x-8)^2}$, select a square window, and press $\boxed{\text{GRAPH}}$. Here we use $[-12, 24, -20, 4]$. The first equation produces the top half of the circle and the second equation produces the lower half.

We can also enter $y_1 = \sqrt{81 - (x-8)^2}$ and then enter $y_2 = -7 + y_1$ and $y_3 = -7 - y_1$. Then deselect y_1, select a square window, and press $\boxed{\text{GRAPH}}$. We use y_1 to eliminate the need to enter the radical expression more than once. Deselecting it prevents the graph of y_1 from appearing on the screen with the graph of the circle. The top half of the graph is produced by y_2 and the lower half by y_3.

```
Plot1  Plot2  Plot3
\Y1◼√(81-(X-8)²)

\Y2◼-7+Y1
\Y3◼-7-Y1■
\Y4=
\Y5=
\Y6=
```

GRAPHING ELLIPSES

The equation of an ellipse must be solved for y before it can be entered on a TI-83, TI-83 Plus, or TI-84 Plus. The procedure for graphing an ellipse of the form $\dfrac{x^2}{a^2} + \dfrac{y^2}{b^2} = 1$ or $\dfrac{x^2}{b^2} + \dfrac{y^2}{a^2} = 1$ is described on page 515 of the text. Here we consider ellipses of the form $\dfrac{(x-h)^2}{a^2} + \dfrac{(y-k)^2}{b^2} = 1$ or $\dfrac{(x-h)^2}{b^2} + \dfrac{(y-k)^2}{a^2} = 1$

Section 6.2, Example 4 Graph the ellipse $4x^2 + y^2 + 24x - 2y + 21 = 0$.

Completing the square in the text, we found that the equation can be written as
$$\frac{(x+3)^2}{4} + \frac{(y-1)^2}{16} = 1.$$
Solve this equation for y.

$$\frac{(x+3)^2}{4} + \frac{(y-1)^2}{16} = 1$$
$$\frac{(y-1)^2}{16} = 1 - \frac{(x+3)^2}{4}$$
$$(y-1)^2 = 16 - 4(x+3)^2 \qquad \text{Multiplying by 16}$$
$$y - 1 = \pm\sqrt{16 - 4(x+3)^2}$$
$$y = 1 \pm \sqrt{16 - 4(x+3)^2}$$

Now we can produce the graph in either of two ways. One is to enter $y_1 = 1 + \sqrt{16 - 4(x+3)^2}$ and $y_2 = 1 - \sqrt{16 - 4(x+3)^2}$, select a square window, and press $\boxed{\text{GRAPH}}$. Here we use $[-9, 9, -6, 6]$. The first equation produces the top half of the ellipse and the second equation produces the lower half.

We can also enter $y_1 = \sqrt{16 - 4(x+3)^2}$ and then enter $y_2 = 1 + y_1$ and $y_3 = 1 - y_1$. Deselect y_1, select a square window, and press $\boxed{\text{GRAPH}}$. We use y_1 to eliminate the need to enter the radical expression more than once. Deselecting it prevents the graph of y_1 from appearing on the screen with the graph of the ellipse. The top half of the graph is produced by y_2 and the lower half by y_3.

We could also begin by using the quadratic formula to solve the original equation for y.

$$4x^2 + y^2 + 24x - 2y + 21 = 0$$
$$y^2 - 2y + (4x^2 + 24x + 21) = 0$$
$$y = \frac{-(-2) \pm \sqrt{(-2)^2 - 4 \cdot 1 \cdot (4x^2 + 24x + 21)}}{2 \cdot 1}$$
$$y = \frac{2 \pm \sqrt{4 - 16x^2 - 96x - 84}}{2}$$
$$y = \frac{2 \pm \sqrt{-16x^2 - 96x - 80}}{2}$$

Then enter $y_1 = \dfrac{2 + \sqrt{-16x^2 - 96x - 80}}{2}$ and $y_2 = \dfrac{2 - \sqrt{-16x^2 - 96x - 80}}{2}$, or enter $y_1 = \sqrt{-16x^2 - 96x - 80}$, $y_2 = \dfrac{2 + y_1}{2}$, and $y_3 = \dfrac{2 - y_1}{2}$, and deselect y_1.

Select a square window and press $\boxed{\text{GRAPH}}$ to display the graph shown above.

GRAPHING HYPERBOLAS

As with equations of circles, parabolas, and ellipses, equations of hyperbolas must be solved for y before they can be entered on a TI-83, TI-83 Plus, or TI-84 Plus.

Section 6.3, Example 2 Graph the hyperbola $9x^2 - 16y^2 = 144$.

First solve the equation for y.

$$9x^2 - 16y^2 = 144$$

$$-16y^2 = -9x^2 + 144$$

$$y^2 = \frac{-9x^2 + 144}{-16}$$

$$y = \pm\sqrt{\frac{-9x^2 + 144}{-16}}, \text{ or } \pm\sqrt{\frac{9x^2 - 144}{16}}$$

It is not necessary to simplify further.

Now enter $y_1 = \sqrt{\frac{9x^2 - 144}{16}}$ and either $y_2 = -\sqrt{\frac{9x^2 - 144}{16}}$ or $y_2 = -y_1$, select a square window, and press $\boxed{\text{GRAPH}}$. Here we use $[-9, 9, -6, 6]$. The top half of the graph is produced by y_1 and the lower half by y_2.

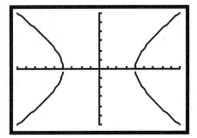

Section 6.3, Example 3 Graph the hyperbola $4y^2 - x^2 + 24y + 4x + 28 = 0$.

In the text we completed the square to get the standard form of the equation. Now solve the equation for y.

$$\frac{(y+3)^2}{1} - \frac{(x-2)^2}{4} = 1$$

$$(y+3)^2 = \frac{(x-2)^2}{4} + 1$$

$$y + 3 = \pm\sqrt{\frac{(x-2)^2}{4} + 1}$$

$$y = -3 \pm \sqrt{\frac{(x-2)^2}{4} + 1}$$

The graph can be produced in either of two ways. One is to enter $y_1 = -3 + \sqrt{\frac{(x-2)^2}{4} + 1}$ and $y_2 = -3 - \sqrt{\frac{(x-2)^2}{4} + 1}$, select a square window, and press $\boxed{\text{GRAPH}}$. Here we use $[-12, 12, -9, 9]$. The first equation produces the top half of the hyperbola and the second the lower half.

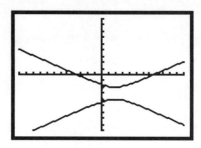

We can also enter $y_1 = \sqrt{\dfrac{(x-2)^2}{4} + 1}$, $y_2 = -3 + y_1$, and $y_3 = -3 - y_1$. Then deselect y_1, select a square window, and press $\boxed{\text{GRAPH}}$. Again, y_1 is used to eliminate the need to enter the radical expression more than once. Deselecting it prevents the graph of y_1 from appearing on the screen with the graph of the hyperbola. The top half of the graph is produced by y_2 and the lower half by y_3.

Chapter 7
Sequences, Series, and Combinatorics

The computational capabilities of a graphing calculator can be used when working with sequences, series, and combinatorics.

FINDING THE TERMS OF A SEQUENCE

Section 7.1, Example 2 Use a graphing calculator to find the first 5 terms of the sequence whose general term is given by $a_n = n/(n+1)$.

Although we could use a table, we will use the Seq feature. The calculator can be set in either Func or Seq mode when this feature is used. Press $\boxed{\text{2nd}}$ $\boxed{\text{LIST}}$ $\boxed{\triangleright}$ to display the LIST OPS menu. Then press 5 to paste "seq(" to the home screen. (LIST is the second operation associated with the $\boxed{\text{STAT}}$ key.) Then enter the general term of the sequence, replacing n with x if the calculator is in Func mode. Follow this with the variable and the numbers of the first and last terms desired. We will also use the ▷Frac feature to express the terms as fractions. Press $\boxed{\text{X, T, } \Theta, n}$ $\boxed{\div}$ $\boxed{(}$ $\boxed{\text{X, T, } \Theta, n}$ $\boxed{+}$ $\boxed{1}$ $\boxed{)}$ $\boxed{,}$ $\boxed{\text{X, T, } \Theta, n}$ $\boxed{,}$ 1 $\boxed{,}$ 5 $\boxed{)}$ $\boxed{\text{MATH}}$ $\boxed{\text{ENTER}}$ $\boxed{\text{ENTER}}$. If the calculator is in Func mode, pressing $\boxed{\text{X, T, } \Theta, n}$ will produce the variable x in the expression. If it is in Sequence mode the variable will be n.

FINDING PARTIAL SUMS

We can use a graphing calculator to find partial sums of a sequence when a formula for the general term is known.

Section 7.1, Example 6 Use a graphing calculator to find S_1, S_2, S_3, and S_4 for the sequence whose general term is given by $a_n = n^2 - 3$.

We will use the cumSum feature. The calculator will write the partial sums as a list. First press $\boxed{\text{2nd}}$ $\boxed{\text{LIST}}$ $\boxed{\triangleright}$ 6 to paste "cumSum(" to the home screen. Then press $\boxed{\text{2nd}}$ $\boxed{\text{LIST}}$ $\boxed{\triangleright}$ 5 to paste "seq(" into the cumSum expression. Finally press $\boxed{\text{X, T, } \Theta, n}$ $\boxed{x^2}$ $\boxed{-}$ 3 $\boxed{,}$ $\boxed{\text{X, T, } \Theta, n}$ $\boxed{,}$ 1 $\boxed{,}$ 4 $\boxed{)}$ $\boxed{)}$ $\boxed{\text{ENTER}}$. We show the result with the calculator set in Seq mode. The variable will be expressed as x if the calculator is in Func mode.

The TI-83, TI-83 Plus, and TI-84 Plus can also compute individual partial sums. This is demonstrated in the next example.

Section 7.1, Example 7 (a) Evaluate $\displaystyle\sum_{k=1}^{5} k^3$.

Press 2nd LIST ▷ ▷ 5 2nd LIST ▷ 5 X, T, Θ, n ∧ 3 , X, T, Θ, n , 1 , 5)) ENTER . We show the result on a calculator set in Seq mode.

RECURSIVELY DEFINED SEQUENCES

Recursively defined sequences can also be entered on a graphing calculator set in Seq mode.

Section 7.1, Example 9 Find the first 5 terms of the sequence defined by

$$a_1 = 5, \ a_{k+1} = 2a_k - 3, \text{ for } k \geq 1.$$

Press Y = and enter the recursive function beside "u(n) =" by pressing 2 2nd u (X,T,θ,n − 1) − 3. (u is the second operation associated with the 7 key.) Also set u(nMin) = 5, the first term of the sequence. Press 2nd TblSet to display the TABLE SETUP screen and set TblStart = 1, ΔTbl = 1, and Indpnt: Auto. Now press 2nd TABLE to display the table of values.

We see that $a_1 = 5$, $a_2 = 7$, $a_3 = 11$, $a_4 = 19$, and $a_5 = 35$.

EVALUATING FACTORIALS, PERMUTATIONS, AND COMBINATIONS

Operations from the MATH PRB (probability) menu can be used to evaluate factorials, permutations, and combinations.

Section 7.5, Exercise 6 Evaluate 7!.

Press 7 | MATH | ▷ | ▷ | ▷ | 4 | ENTER |. These keystrokes enter 7, display the MATH PRB menu, select item 4, !, from that menu, and then cause 7! to be evaluated. The result is 5040.

Section 7.5, Exercise 9 Evaluate $\dfrac{9!}{5!}$.

Press 9 | MATH | ▷ | ▷ | ▷ | 4 | ÷ | 5 | MATH | ▷ | ▷ | ▷ | 4 | ENTER |. The result is 3024.

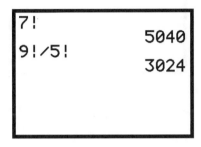

Section 7.5, Example 3 (a) Compute $_4P_4$.

Press 4 | MATH | ▷ | ▷ | ▷ | 2 4 | ENTER |. These keystrokes enter 4, for 4 objects, display the MATH PRB menu, select item 2, $_nP_r$, from that menu, enter 4, for 4 objects taken at a time, and then cause the calculation to be performed. The result is 24.

Section 7.5, Example 6 Compute $_8P_4$.

Press 8 | MATH | ▷ | ▷ | ▷ | 2 4 | ENTER |. These keystrokes enter 8, for 8 objects, display the MATH PRB menu, select item 2, $_nP_r$, from that menu, enter 4, for 4 objects taken at a time, and then cause the calculation to be performed. The result is 1680.

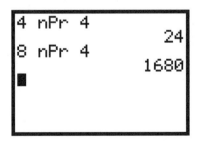

Section 7.6, Example 2 Evaluate $\begin{pmatrix} 7 \\ 5 \end{pmatrix}$.

Press 7 $\boxed{\text{MATH}}$ $\boxed{\triangleright}$ $\boxed{\triangleright}$ $\boxed{\triangleright}$ 3 5 $\boxed{\text{ENTER}}$. These keystrokes enter 7, for 7 objects, display the MATH PRB menu, select item 3, $_nC_r$, from that menu, enter 5, for 5 objects taken at a time, and then cause the calculation to be performed. The result is 21.

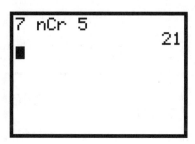

The TI-89 Graphics Calculator

Chapter R
Basic Concepts of Algebra

GETTING STARTED

Before turning on the calculator note that there are options above the keys as well as on them. To access the option written on a key, simply press the key. The options written in yellow above the keys are accessed by first pressing the yellow $\boxed{\text{2nd}}$ key in the left column of the keypad and then pressing the key corresponding to the desired option. The purple options above the keys are accessed by first pressing the purple $\boxed{\text{alpha}}$ key in the second column of the keypad. The green options are accessed by first pressing the $\boxed{\diamond}$ key in the left column of the keypad. This key has a green diamond inside a green border.

Press $\boxed{\text{ON}}$ to turn on the TI-89 graphing calculator. ($\boxed{\text{ON}}$ is the key at the bottom left-hand corner of the keypad.) The home screen is displayed. You should see a row of boxes at the top of the screen and two horizontal lines with lettering below them at the bottom of the screen. If you do not see anything, try adjusting the display contrast. To do this, first press and hold down the $\boxed{\diamond}$ key. Then press and hold $\boxed{+}$ to darken the display or $\boxed{-}$ to lighten the display. Be sure to use the black $\boxed{-}$ key in the right column of the keypad rather than the gray $\boxed{(-)}$ key on the bottom row.

One way to turn the calculator off is to press $\boxed{\text{2nd}}$ $\boxed{\text{OFF}}$. (OFF is the second operation associated with the $\boxed{\text{ON}}$ key.) When you turn the TI-89 on again the home screen will be displayed regardless of the screen that was displayed when the calculator was turned off. $\boxed{\text{2nd}}$ $\boxed{\text{OFF}}$ cannot be used to turn off the calculator if an error message is displayed. The calculator can also be turned off by pressing $\boxed{\diamond}$ $\boxed{\text{OFF}}$. This will work even if an error message is displayed. After $\boxed{\diamond}$ $\boxed{\text{OFF}}$ is used, when the TI-89 is turned on again the display will be exactly as it was when it was turned off. The calculator will turn itself off automatically after several minutes without any activity. When this happens the display will be just as you left it when you turn the calculator on again.

From top to bottom, the home screen consists of the toolbar, the large history area where entries and their corresponding results are displayed, the entry line where expressions or instructions are entered, and the status line which shows the current state of the calculator. These areas will be discussed in more detail as the need arises.

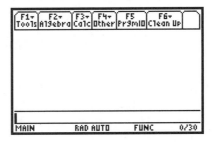

Press $\boxed{\text{MODE}}$ to display the MODE settings. Modes that are not currently valid, due to the existing choices of settings, are dimmed. Initially you should select the settings shown below.

To change a setting on the Mode screen use $\boxed{\triangledown}$ or $\boxed{\triangle}$ to move the cursor to the line of that setting. Then use $\boxed{\triangleright}$ to display the options. Press the number of the desired option to copy it to the Mode screen. Then press $\boxed{\text{ENTER}}$ to save it there. Instead of pressing the number of the desired setting, you can highlight it and then press $\boxed{\text{ENTER}}$ $\boxed{\text{ENTER}}$ to copy it to the Mode screen and save it there. Note that the cursor skips dimmed settings as you move through the options.

It will be helpful to read Chapter 1: Getting Started and Chapter 2: Operating the TI-89 in the TI-89 Guidebook before proceeding.

USING A MENU

A menu is a list of options that appear when a key is pressed. For example, press $\boxed{\text{F1}}$ to display the Tools menu. We can select an item from a menu by using $\boxed{\triangledown}$ to highlight it and then pressing $\boxed{\text{ENTER}}$ or by simply pressing the number of the item. If we press 8 when the Tools menu is displayed, for instance, we select "Clear Home" and any previously entered computations will be cleared from the history area of the home screen. If an item is identified by a letter rather than a number, press the purple $\boxed{\text{alpha}}$ key followed by the letter of the item to select it. The letters are printed in purple above the keys on the keypad. The down-arrow beside item 8 in the menu below indicates that there are additional items in the menu. Use $\boxed{\triangledown}$ to scroll down to them.

ABSOLUTE VALUE

Section R.1, Example 3 Find the distance between -2 and 3.

First press $\boxed{\text{HOME}}$ or $\boxed{\text{2nd}}$ $\boxed{\text{QUIT}}$ to go to the home screen. You might want to clear any previously entered computations from the history area of the home screen first. To do this, access Tools from the toolbar at the top of the screen by pressing $\boxed{\text{F1}}$, the blue key at the top left-hand corner of the keypad. Then select item 8, Clear Home, from this menu by pressing 8.

The entry line on the home screen can be cleared by pressing $\boxed{\text{CLEAR}}$. This is not necessary if the current entry is highlighted, since it will automatically be cleared when the first character of a new entry is entered.

The distance between -2 and 3 is $|-2-3|$, or $|3-(-2)|$. Absolute value notation is denoted "abs" on the TI-89. It is item 2 on the MATH Number menu. Note that the gray $\boxed{(-)}$ key in the bottom row of the keypad must be used to enter a negative number. The black $\boxed{-}$ key in the right column of the keypad is used for the subtraction operation.

To enter $|-2-3|$ press $\boxed{\text{2nd}}$ $\boxed{\text{MATH}}$ $\boxed{\triangleright}$ $\boxed{2}$ $\boxed{(-)}$ $\boxed{2}$ $\boxed{-}$ $\boxed{3}$ $\boxed{)}$ $\boxed{\text{ENTER}}$. (MATH is the second operation associated with the 5 numeric key.)To enter $|3-(-2)|$ press $\boxed{\text{2nd}}$ $\boxed{\text{MATH}}$ $\boxed{\triangleright}$ $\boxed{2}$ 3 $\boxed{-}$ $\boxed{(}$ $\boxed{(-)}$ $\boxed{2}$ $\boxed{)}$ $\boxed{)}$ $\boxed{\text{ENTER}}$. Note that the calculator supplies the left parenthesis in the absolute value notation. We close the expression with a right parenthesis. Although the parentheses around -2 in the second expression are not necessary, they allow the expression to be more easily read on the entry line so we include them here.

Instead of pressing $\boxed{\text{2nd}}$ $\boxed{\text{MATH}}$ $\boxed{\triangleright}$ 2 to access "abs(" and copy it to the home screen, we could have pressed $\boxed{\text{2nd}}$ $\boxed{\text{MATH}}$ $\boxed{\triangleright}$ $\boxed{\triangledown}$ $\boxed{\text{ENTER}}$. Absolute value notation can also be found as the first item in the Catalog and copied to the home screen. Access the Catalog by pressing $\boxed{\text{CATALOG}}$. Then use the $\boxed{\triangle}$ or $\boxed{\triangledown}$ key to position the triangular selection cursor beside "abs(." This cursor can be positioned quickly by pressing $\boxed{\text{A}}$ to move the cursor to the first item in the Catalog that begins with A 'abs(." (A is the purple alphabetic operation associated with the $\boxed{=}$ key.) Note that, when we are in the Catalog, it is not necessary to press $\boxed{\text{alpha}}$ before an alphabetic entry.

EDITING ENTRIES

You can edit your entry if necessary. After $\boxed{\text{ENTER}}$ is pressed to evaluate an expression, the TI-89 leaves the expression on the entry line and highlights it. To edit the expression you must first remove the highlight to avoid the possibility of accidently typing over the entire expression. To do this, press $\boxed{\triangleleft}$ or $\boxed{\triangleright}$ to move the cursor (a blinking vertical line) toward the side of the expression to be edited. If, for instance, in entering one of the expressions in Example 3 above you pressed 6 instead of 3, first press $\boxed{\triangleleft}$ to move the cursor to the beginning of the expression or $\boxed{\triangleright}$ to move it to the end of the expression. Now, to type a 6 over the 3, first select overtype mode by pressing $\boxed{\text{2nd}}$ $\boxed{\text{INS}}$. (INS is the second operation associated with the $\boxed{\leftarrow}$ key.) Now the cursor becomes a dark, blinking rectangle rather than a vertical line. Use $\boxed{\triangleright}$ to position the cursor over the 6 and then press 3. To leave overtype mode press $\boxed{\text{2nd}}$ $\boxed{\text{INS}}$ again. The calculator is now in the insert mode, indicated by a vertical cursor, and will remain in that mode until overtype mode is once again selected.

If you forgot to type the 2 in the first expression, move the insert cursor to the left of the subtraction symbol and press 2 to insert the 2 before that symbol. You can continue to insert symbols immediately after the first insertion. If you typed 21 instead of 2, move the cursor to the left of 1 and press $\boxed{\leftarrow}$. This will delete the 1. Instead of using overtype mode to overtype a character as described above, we can use $\boxed{\leftarrow}$ to delete the character and then, in insert mode, insert a new character.

If you accidently press $\boxed{\triangle}$ instead of $\boxed{\triangleleft}$ or $\boxed{\triangleright}$ while editing an expression, the cursor will move up into the history area of the screen. Press $\boxed{\text{ESC}}$ to return immediately to the entry line. The $\boxed{\triangledown}$ key can also be used to return to the entry line. It must be pressed the same number of times the $\boxed{\triangle}$ key was pressed accidently.

If you notice that an entry needs to be edited before you press $\boxed{\text{ENTER}}$ to perform the computation, the editing can be done as described above without the necessity of first removing the highlight from the entry.

The keystrokes $\boxed{\text{2nd}}$ $\boxed{\text{ENTRY}}$ can be used repeatedly to recall entries preceding the last one. (ENTRY is the second function associated with the ENTER key.) Pressing 2nd ENTRY twice, for example, will recall the next to last entry. Using these keystrokes a third time recalls the third to last entry and so on. The number of entries that can be recalled depends on the amount of storage they occupy in the calculator's memory.

Previous entries and results of computations can also be copied to the entry line by first using the △ key to move through the history area until the desired entry or result is highlighted. Then press ENTER to copy it to the entry line.

SCIENTIFIC NOTATION

To enter a number in scientific notation, first type the decimal portion of the number; then press the EE key in the left column of the keypad; finally type the exponent, which can be at most three digits. For example, to enter 1.789×10^{-11} in scientific notation, go to the home screen and press 1 . 7 8 9 EE (−) 1 1 ENTER. To enter 6.084×10^{23} in scientific notation, press 6 . 0 8 4 EE 2 3 ENTER. The decimal portion of each number appears before a small E while the exponent follows the E.

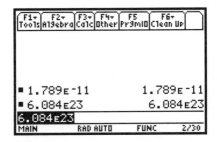

The TI-89 can be used to perform computations in scientific notation.

Section R.2, Example 8 *Distance to a Star.* The nearest star, Alpha Centauri C, is about 4.22 light-years from Earth. One light-year is the distance that light travels in one year and is about 5.88×10^{12} miles. How many miles is it from Earth to Alpha Centauri? Express your answer in scientific notation.

To solve this problem we find the product $4.22 \times (5.88 \times 10^{12})$. Press 4 . 2 2 × 5 . 8 8 2nd EE 1 2 ENTER. The result is 2.48136×10^{13} miles.

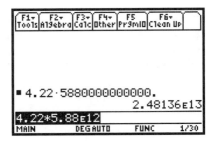

ORDER OF OPERATIONS

Section R.2, Example 9 (b) Calculate: $\dfrac{10 \div (8-6) + 9 \cdot 4}{2^5 + 3^2}$.

In order to divide the entire numerator by the entire denominator, we must enclose both the numerator and the denominator in parentheses. That is, we enter $(10 \div (8-6) + 9 \cdot 4) \div (2^5 + 3^2)$. On the home screen press $\boxed{(}$ $1\,0$ $\boxed{\div}$ $\boxed{(}$ 8 $\boxed{-}$ 6 $\boxed{)}$ $\boxed{+}$ 9 $\boxed{\times}$ 4 $\boxed{)}$ $\boxed{\div}$ $\boxed{(}$ 2 $\boxed{\wedge}$ 5 $\boxed{+}$ 3 $\boxed{\wedge}$ 2 $\boxed{)}$ $\boxed{\text{ENTER}}$.

RADICAL NOTATION

We can use the square-root key and rational exponents to simplify radical expressions.

Section R.6, Example 1 Simplify each of the following.

a) $\sqrt{36}$ **b)** $-\sqrt{36}$ **c)** $\sqrt[5]{\dfrac{32}{243}}$ **d)** $\sqrt[3]{-8}$ **e)** $\sqrt[4]{-16}$

a) To find $\sqrt{36}$ go to the home screen and press $\boxed{\text{2nd}}$ $\boxed{\sqrt{}}$ $3\,6$ $\boxed{)}$ $\boxed{\text{ENTER}}$. ($\sqrt{}$ is the second operation associated with the $\boxed{\times}$ multiplication key.) Note that the calculator supplies a left parenthesis with the radical symbol and we must close the expression with a right parenthesis.

b) To find $-\sqrt{36}$ press $\boxed{(-)}$ $\boxed{\text{2nd}}$ $\boxed{\sqrt{}}$ $3\,6$ $\boxed{)}$ $\boxed{\text{ENTER}}$.

c) We will use a rational exponent to find $\sqrt[5]{\dfrac{32}{243}}$. We have $\sqrt[5]{\dfrac{32}{243}} = \left(\dfrac{32}{243}\right)^{1/5}$. Press $\boxed{(}$ $3\,2$ $\boxed{\div}$ $2\,4\,3$ $\boxed{)}$ $\boxed{\wedge}$ $\boxed{(}$ 1 $\boxed{\div}$ 5 $\boxed{)}$ $\boxed{\text{ENTER}}$.

d) We will also use a rational exponent to find $\sqrt[3]{-8}$. We have $\sqrt[3]{-8} = (-8)^{1/3}$. Press $\boxed{(}$ $\boxed{(-)}$ 8 $\boxed{)}$ $\boxed{\wedge}$ $\boxed{(}$ 1 $\boxed{\div}$ 3 $\boxed{)}$ $\boxed{\text{ENTER}}$.

e) We have $\sqrt[4]{-16} = (-16)^{1/4}$. Press $\boxed{(}$ $\boxed{(-)}$ $\boxed{1}$ $\boxed{6}$ $\boxed{)}$ $\boxed{\wedge}$ $\boxed{(}$ $\boxed{1}$ $\boxed{\div}$ $\boxed{4}$ $\boxed{)}$ $\boxed{\text{ENTER}}$. When Real is selected for Complex Format on the MODE screen, the calculator returns an error message indicating that the result is not a real number.

Chapter 1
Graphs, Functions, and Models

GRAPHING EQUATIONS

Section 1.1, Example 4 Graph $3x - 5y = -10$.

First we solve for y as shown in the text, obtaining $y = \dfrac{3}{5}x + 2$. Then press $\boxed{\diamond}$ $\boxed{\text{Y} =}$ to access the equation-editor screen. (Y = is the green $\diamond$ operation associated with the $\boxed{\text{F1}}$ key.) If any plots are turned on they should be turned off, or deselected, now. A check mark beside the name of a plot indicates that it is currently selected. To deselect it, move the cursor to the plot. Then press $\boxed{\text{F4}}$. There should now be no check mark beside the name of the plot, indicating that it has been deselected. If there is currently an expression displayed for y_1, clear it by positioning the cursor beside "y1 =" and then press $\boxed{\text{CLEAR}}$. Do the same for expressions that appear on all other "y =" lines by using $\boxed{\triangledown}$ to move to a line and then pressing $\boxed{\text{CLEAR}}$. Then use $\boxed{\triangle}$ or $\boxed{\triangledown}$ to move the cursor beside "y1 =." Now enter $y_1 = \dfrac{3}{5}x + 2$ on the entry line of the equation-editor screen and paste it beside "y1 =" by pressing $\boxed{(}$ 3 $\boxed{\div}$ 5 $\boxed{)}$ $\boxed{\text{X}}$ $\boxed{+}$ 2 $\boxed{\text{ENTER}}$. Although the parentheses are not necessary, the equation is more easily read on the entry line when they are used.

Next we set the viewing window. This is the portion of the coordinate plane that appears on the calculator's screen. It is defined by the minimum and maximum values of x and y: xmin, xmax, ymin, and ymax. The notation [xmin, xmax, ymin, ymax] is used in the text to represent these window settings or dimensions. For example, $[-12, 12, -8, 8]$ denotes a window that displays the portion of the x-axis from -12 to 12 and the portion of the y-axis from -8 to 8. In addition, the distance between tick marks on the axes is defined by the settings xscl and yscl. In this manual xscl and yscl will be assumed to be 1 unless noted otherwise. The setting xres sets the pixel resolution. We usually select xres = 2. The window corresponding to the settings $[-20, 30, -12, 20]$, xscl = 5, yscl = 2, xres = 2, is shown below.

Press ⬜◇⬜ ⬜WINDOW⬜ to display the current window settings on your calculator. (WINDOW is the green ◇ operation associated with the ⬜F2⬜ key on the top row of the keypad.) The standard settings are shown below.

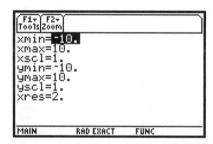

To change a setting, position the cursor beside the setting you wish to change and enter the new value. For example, to change from the standard settings to $[-20, 30, -12, 20]$, xscl = 5, yscl = 2, on the WINDOW screen, start with the setting beside "xmin =" highlighted and press ⬜(−)⬜ 2 0 ⬜ENTER⬜ 3 0 ⬜ENTER⬜ 5 ⬜ENTER⬜ ⬜(−)⬜ 1 2 ⬜ENTER⬜ 2 0 ⬜ENTER⬜ 2 ⬜ENTER⬜. You must use the gray ⬜(−)⬜ key on the bottom row of the keypad rather than the black ⬜−⬜ key in the right-hand column to enter a negative number. ⬜(−)⬜ represents "the opposite of" or "the additive inverse of" whereas ⬜−⬜ is the key for the subtraction operation. The ⬜▽⬜ key may be used instead of ⬜ENTER⬜ after typing each window setting. To see the window $[-20, 30, -12, 20]$, xscl = 5, yscl = 2 shown above, press ⬜◇⬜ ⬜GRAPH⬜. (GRAPH is the green ◇ operation associated with the ⬜F3⬜ key on the top row of the keypad.)

QUICK TIP: To return quickly to the standard window setting $[-10, 10, -10, 10]$, xscl = 1, yscl = 1, when either the Window screen or the Graph screen is displayed, press ⬜F2⬜ to access the ZOOM menu and then press 6 to select item 6, ZoomStd (Zoom Standard).

The standard window is a good choice for the graph of the equation $y = \frac{3}{5}x + 2$. Either enter these dimensions in the WINDOW screen and then press ⬜◇⬜ ⬜GRAPH⬜ to see the graph or, from the WINDOW screen, simply press ⬜F2⬜ 6 to select the standard window and see the graph.

THE TABLE FEATURE

A table of x-and y-values representing ordered pairs that are solutions of an equation can be displayed on the TI-89.

Section 1.1, Example 5 Graph: $y = x^2 - 9x - 12$.

If a hand-drawn graph of an equation is desired, a graphing calculator can be used to create a table of ordered pairs that are solutions of the equation. These ordered pairs can then be plotted and connected with a smooth curve to graph the equation. To create such a table for the equation $y = x^2 - 9x - 12$, we first enter the equation on the equation-editor screen. Once the equation in entered, press $\boxed{\diamond}$ $\boxed{\text{TblSet}}$ or $\boxed{\diamond}$ $\boxed{\text{TABLE}}$ $\boxed{\text{F2}}$ to access the TABLE SETUP window. (TblSet is the green $\diamond$ operation associated with the $\boxed{\text{F4}}$ key.) If "Independent" is set to "Auto" on the Table Setup screen, the calculator will supply values for x, beginning with the value specified as tblStart and continuing by adding the value of Δtbl to the preceding value for x. If the table was previously set to Ask, the blinking cursor will be positioned over ASK. Change this setting to AUTO by pressing $\boxed{\triangleright}$ 1. Now use the $\boxed{\triangle}$ key to move the cursor to tblStart. Enter a minimum x-value of -3, an increment of 1, and a Graph $< - >$ Table setting of OFF by first positioning the cursor beside tblStart and then pressing $\boxed{(-)}$ 3 $\boxed{\triangledown}$ 1 $\boxed{\triangledown}$ $\boxed{\triangleright}$ 1 $\boxed{\text{ENTER}}$. Press $\boxed{\diamond}$ $\boxed{\text{TABLE}}$ to see the table. (TABLE is the green $\diamond$ operation associated with the $\boxed{\text{F5}}$ key.) We can use the $\boxed{\triangle}$ and $\boxed{\triangledown}$ keys to scroll up and down in the table to find values other than those shown here.

SQUARING THE VIEWING WINDOW

In the standard window, the distance between tick marks on the y-axis is about $1/2$ the distance between tick marks on the x-axis. It is often desirable to choose window dimensions for which these distances are the same, creating a "square" window. Any window in which the ratio of the length of the y-axis to the length of the x-axis is $1/2$ will produce this effect.

This can be accomplished by selecting dimensions for which $\text{ymax} - \text{ymin} = \frac{1}{2}(\text{xmax} - \text{xmin})$. For example, the windows $[-12, 12, -6, 6]$ and $[-6, 6, -3, 3]$ are square. To illustrate this, we graph the circle $x^2 + y^2 = 9$ in the standard window. Note that the graph does not appear to be a circle. (We will explain how to graph a circle later in this chapter.)

Now change the window dimensions to $[-8, 8, -4, 4]$, xscl $= 1$, yscl $= 1$, and press $\boxed{\diamond}$ $\boxed{\text{GRAPH}}$. Observe that the distance between tick marks appears to be the same on both axes and that the graph appears to be a circle.

The window can also be squared using the calculator's ZoomSqr feature. From the equation-editor, Window, or Graph screen, press $\boxed{\text{F2}}$ 5 to select the ZoomSqr window. The resulting window dimensions and graph are shown below. Here the window was squared from the standard window. Note that the graph also appears to be a circle in this window.

GRAPHING CIRCLES

If the center and radius of a circle are known, the circle can be graphed using the Circle feature from the Catalog.

Section 1.1, Example 11 Graph $(x + 5)^2 + (y - 2)^2 = 16$.

The center of this circle is $(-5, 2)$ and its radius is 4. To graph it first press $\boxed{\diamond}$ $\boxed{Y =}$ and clear all previously entered equations on the equation-editor screen. Then select a square window. The dimensions $[-14, 6, -3, 7]$, xscl $= 2$, are a good choice for this circle. Now, to select Circle from the Catalog, first press $\boxed{\text{HOME}}$ or $\boxed{\text{2nd}}$ $\boxed{\text{QUIT}}$ to go to the home screen. Then press $\boxed{\text{CATALOG}}$ $\boxed{\text{C}}$ to go to the beginning of the items in the Catalog that start with C. (Note that it is not necessary to press $\boxed{\text{alpha}}$ before a letter key when the Catalog is displayed.) Now use $\boxed{\triangledown}$ to go to "Circle" and press $\boxed{\text{ENTER}}$. The Circle command appears on the entry line of the home screen. "Circle" can also be typed directly on the entry line of the home screen by pressing $\boxed{\text{2nd}}$ $\boxed{\text{a-lock}}$ $\boxed{\text{C}}$ $\boxed{\text{I}}$ $\boxed{\text{R}}$ $\boxed{\text{C}}$ $\boxed{\text{L}}$ $\boxed{\text{E}}$ $\boxed{\text{alpha}}$. Enter the x-coordinate of the center, the y-coordinate of the center, and the length of the radius, all separated by commas. To do this press $\boxed{(-)}$ 5 $\boxed{,}$ 2 $\boxed{,}$ 4. Press $\boxed{\text{ENTER}}$ to see the graph.

This graph can be cleared from the Graph screen by pressing $\boxed{\text{F4}}$ (ReGraph) or by pressing $\boxed{\text{2nd}}$ $\boxed{\text{F6}}$ to display the Draw menu and then pressing 1 to select ClrDraw. The ClrDraw command can also be accessed from the Catalog. From the home screen, press $\boxed{\text{CATALOG}}$ $\boxed{\text{C}}$, scroll to ClrDraw, and press $\boxed{\text{ENTER}}$ $\boxed{\text{ENTER}}$.

FINDING FUNCTION VALUES

When a formula for a function is given, function values can be found in several ways.

Section 1.2, Example 4 (b) For $f(x) = 2x^2 - x + 3$, find $f(-7)$.

Method 1: Substitute the inputs directly in the formula. On the home screen press 2 $\boxed{(}$ $\boxed{(-)}$ 7 $\boxed{)}$ $\boxed{\wedge}$ 2 $\boxed{-}$ $\boxed{(}$ $\boxed{(-)}$ 7 $\boxed{)}$ $\boxed{+}$ 3 $\boxed{\text{ENTER}}$.

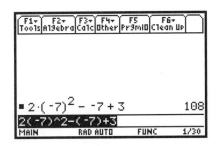

Method 2: Enter $y_1 = 2x^2 - x + 3$ on the "y =" screen. Then press $\boxed{\text{HOME}}$ or $\boxed{\text{2nd}}$ $\boxed{\text{QUIT}}$ to go to the home screen. To find $f(-7)$, the value of y_1 when $x = -7$, press $\boxed{(-)}$ 7 $\boxed{\text{STO} \triangleright}$ $\boxed{\text{X}}$ $\boxed{\text{2nd}}$ $\boxed{:}$ $\boxed{\text{Y}}$ 1 $\boxed{(}$ $\boxed{\text{X}}$ $\boxed{)}$ $\boxed{\text{ENTER}}$. (: is the second operation associated with the 4 numeric key.) This series of keystrokes stores -7 as the value of x and then substitutes it in the function y_1.

Method 3: Enter $y_1 = 2x^2 - x + 3$ on the "Y =" screen and press $\boxed{\text{HOME}}$ or $\boxed{\text{2nd}}$ $\boxed{\text{QUIT}}$ to go to the home screen. Then press $\boxed{\text{Y}}$ 1 $\boxed{(}$ $\boxed{(-)}$ 7 $\boxed{)}$ $\boxed{\text{ENTER}}$. Note that this entry closely resembles function notation.

Method 4: The TABLE feature can also be used to find function values. Enter $y_1 = 2x^2 - x + 3$ on the "Y =" screen. Then set up a table Ask mode. Press $\boxed{\diamond}$ $\boxed{\text{TblSet}}$ or $\boxed{\diamond}$ $\boxed{\text{TABLE}}$ $\boxed{\text{F2}}$ to access the TableSetup screen. Move the cursor to the "Independent" line. Then press $\boxed{\triangleright}$ 2 $\boxed{\text{ENTER}}$ to select Ask mode. In Ask mode the calculator disregards the other settings on the Table Setup screen.

Now press $\boxed{\diamond}$ $\boxed{\text{TABLE}}$ to view the table. (TABLE is the second operation associated with the $\boxed{\text{F5}}$ key.) If you select Ask before a table is displayed for the first time on your calculator, a blank table is displayed. If a table has previously been displayed, the table you now see will continue to show the values in the previous table.

Values for x can be entered in the x-column of the table and the corresponding values for y_1 will be displayed in the y_1-column. To enter -7, press $\boxed{(-)}$ 7 $\boxed{\text{ENTER}}$. Any additional x-values that are displayed are from a table that was previously displayed on the Auto setting. We see that $y_1 = 108$ when $x = -7$, so $f(-7) = 108$.

Method 5: We can also use the Value feature from the Math menu on the GRAPH screen to find $f(-7)$. To do this, graph $y_1 = 2x^2 - x + 3$ in a window that includes the x-value -7. We will use the standard window. Then press $\boxed{\text{F5}}$ $\boxed{\text{ENTER}}$ to access the Math menu and select item 1, Value. Now supply the desired x-value by pressing $\boxed{(-)}$ 7. Press $\boxed{\text{ENTER}}$ to see $x = -7$, $y = 108$ at the bottom of the screen, Thus, $f(-7) = 108$.

GRAPHS OF FUNCTIONS

The TI-89 does not use function notation. To graph a function, first replace the function notation with y. For example, to graph $f(x) = x^2 - 5$ replace $f(x)$ with y. Then enter the equation $y = x^2 - 5$ on the equation-editor screen and graph it as described on page 67 of this manual.

LINEAR REGRESSION

We can use the Linear Regression feature to fit a linear equation to a set of data.

Section 1.4, Example 10 The following table shows the number of U. S. travelers to foreign countries, in millions, for years since 1994. Fit a regression line to the data. Then use the function to estimate the number of U. S. travelers to foreign countries in 2006.

Years, x	Number of U. S. Travelers to Foreign Countries (in millions)
1994, 0	4.65
1995, 1	5.08
1996, 2	5.23
1997, 3	5.29
1998, 4	5.63
1999, 5	5.75
2000, 6	6.08

We will enter the data as ordered pairs in the Data/Matrix editor. Press $\boxed{\text{APPS}}$ 6 3 to display a new data variable screen in the Data/Matrix editor. We must now enter a data variable name in the Variable box on this screen. The name can contain from 1 to 8 characters and cannot start with a numeral. Some names are preassigned to other uses on the TI-89. If you try to use one of these, you will get an error message. Press $\boxed{\bigtriangledown}$ $\boxed{\bigtriangledown}$ to move the cursor to the Variable box. We will name our data variable "travel." To enter this name, first lock the alphabetic keys on by pressing $\boxed{\text{2nd}}$ $\boxed{\text{a-lock}}$. Then press $\boxed{\text{T}}$ $\boxed{\text{R}}$ $\boxed{\text{A}}$ $\boxed{\text{V}}$ $\boxed{\text{E}}$ $\boxed{\text{L}}$. The letter T has its own key on the sixth row from the bottom of the keypad. The letters R, A, V, E, and L are the purple alphabetic operations associated with the 2, $\boxed{=}$, 0, $\boxed{\div}$, and 4 keys, respectively.

After typing the name of the data variable, unlock the alphabetic keys by pressing the purple $\boxed{\text{alpha}}$ key. Now press $\boxed{\text{ENTER}}$ $\boxed{\text{ENTER}}$ to go to the data-entry screen. Assuming the data variable name "travels" has not previously been used in your calculator, this screen will contain empty data lists with row 1, column 1 highlighted. If entries have previously been made in a data variable named "travel," they can be cleared by pressing $\boxed{\text{F1}}$ 8 $\boxed{\text{ENTER}}$.

We will enter the first coordinates of the points as the number of years since 1994 in column c1 and the second coordinates in c2. To enter 0 press 0 $\boxed{\text{ENTER}}$. Continue typing the first coordinates 1, 2, 3, 4, 5, and 6, each followed by $\boxed{\text{ENTER}}$. The entries can be followed by $\boxed{\bigtriangledown}$ rather than $\boxed{\text{ENTER}}$ if desired. Press $\boxed{\triangleright}$ $\boxed{\triangle}$ $\boxed{\triangle}$ $\boxed{\triangle}$ $\boxed{\triangle}$ $\boxed{\triangle}$ $\boxed{\triangle}$ to move to the top of column c2. Type the second coordinates 4.65, 5.08, 5.23, 5.29, 5.63, 5.75, and 6.08 in succession, each followed by $\boxed{\text{ENTER}}$ or $\boxed{\bigtriangledown}$. Note that the coordinates of each point must be in the same position in both lists.

The calculator can then plot these points, creating a scatterplot of the data. To do this we first access the Plot Setup screen by pressing F2. We will use Plot 1, which is highlighted. If any plot settings are currently entered beside "Plot 1," clear them by pressing F3. Clear settings shown beside any other plots as well by using ▽ to highlight each plot in turn and then pressing F3.

Now we define Plot 1. Use △ to highlight Plot 1 if necessary. Then press F1 to display the Plot Definition screen. The item on the first line, Plot Type, is highlighted. We will choose a scatterplot, by pressing ▷ 1. Now press ▽ to go to the next line, Mark. Here we select the type of mark or symbol that will be used to plot the points. We select a box by pressing ▷ 1. Now we must tell the calculator which columns of the data variable to use for the x- and y-coordinates of the points to be plotted. Press ▽ to move the cursor to the "x" line and enter c1 as the source of the x-coordinates by pressing alpha C 1. (C is the purple alphabetic operation associated with the) key.) Press ▽ alpha C 2 to go the the "y" line and enter c2 as the source of the y-coordinates.

Save the plot definition and return to the Plot Setup screen by pressing ENTER ENTER. Beside "Plot 1:" you will now see a shorthand notation for the definition of the plot. The check mark to the left of Plot 1 indicates that it is turned on.

Note that there should be no equations entered on the equation-editor screen. If there are equations present, clear them as described on page 67 of this manual.

Now we select a window. The x-values range from 0 through 6 and the y-values from 4.65 through 6.08, so one good choice might be $[-2, 7, 0, 7]$. We can enter these dimensions and then press ◇ GRAPH to see the scatterplot. Instead of entering the window dimensions directly, we can press F2 9 after entering the coordinates of the points in lists and defining Plot 1. This activates the ZoomData operation which automatically defines a viewing window that displays all the points and also displays the graph. We use ZoomData to produce the scatterplot shown below.

To turn off the plot, highlight Plot 1 on the equation-editor screen and then press F4 . Note that there is no check mark to the left of Plot 1 when it is turned off.

Now we will fit a linear equation to the data. Press F5 to display the Calculate menu. Press ▷ 5 to select LinReg (linear regression). Then press ▽ alpha C 1 ▽ alpha C 2 to indicate that the data in c1 and c2 will be used for x and y, respectively. Press ▽ ▷ ▽ ENTER to indicate that the regression equation should be copied to the equation-editor screen as y_1. Finally press ENTER again to see the STAT VARS screen which displays the coefficients a and b of the regression equation $y = ax + b$.

Note that values for "corr" (the correlation coefficient) and r^2 (the coefficient of determination) will also be displayed. These numbers indicate how well the regression line fits the data. While it is possible to suppress these numbers on some graphing calculators, this cannot be done on the TI-89.

To estimate the number of U. S. travelers to foreign countries in 2006, evaluate the regression equation for $x = 12$. (2006 is 12 years after 1994.) Use any of the methods for evaluating a function presented earlier in this chapter. (See pages 71 - 73 of this manual.) We will use function notation on the home screen. When $x = 12, y \approx 7.33$, so we predict that there will be about 7.33 million U. S. travelers to foreign countries in 2006.

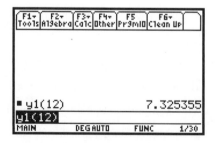

We can also plot the data points along with the graph of the regression equation. To do this we first define a plot as described on page 75 of this manual. Now select a viewing window. Press $\boxed{\diamond}$ $\boxed{\text{WINDOW}}$ to go to the WINDOW screen and then press $\boxed{\text{F2}}$ 9 to activate the ZoomData operation which automatically defines a viewing window that displays all of the points and also displays the graph.

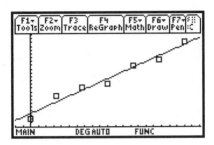

Turn off the plot as described on page 76 of this manual before graphing other functions.

THE MAXIMUM AND MINIMUM FEATURES

Section 1.5, Example 2   Use a graphing calculator to determine any relative maxima or minima of the function $f(x) = 0.1x^3 - 0.6x^2 - 0.1x + 2$.

First graph $y_1 = 0.1x^3 - 0.6x^2 - 0.1x + 2$ in a window that displays the relative extrema of the function. Trial and error reveals that one good choice is $[-4, 6, -3, 3]$. Observe that a relative maximum occurs near $x = 0$ and a relative minimum occurs near $x = 4$.

To find the relative maximum, first press $\boxed{\text{F5}}$ 4 or $\boxed{\text{F5}}$ $\boxed{\triangledown}$ $\boxed{\triangledown}$ $\boxed{\triangledown}$ $\boxed{\text{ENTER}}$ to select the Maximum feature from the Math menu on the Graph screen. We are prompted to select a lower bound for the relative maximum. This means that we must choose an x-value that is to the left of the x-value of the point where the relative maximum occurs. This can be done by using the left- and right-arrow keys to move the cursor to a point to the left of the relative maximum or by keying in an appropriate value.

Once this is done, press $\boxed{\text{ENTER}}$. Now we are prompted to select an upper bound. We move the cursor to a point to the right of the relative maximum or we key in an appropriate value.

Press $\boxed{\text{ENTER}}$ again. We see that a relative maximum function value of approximately 2.004 occurs when $x \approx -0.082$.

To find the relative minimum, select the Minimum feature from the Math menu by pressing $\boxed{\text{F5}}$ 3 or $\boxed{\text{F5}}$ $\boxed{\triangledown}$ $\boxed{\triangledown}$ $\boxed{\text{ENTER}}$. Select lower and upper bounds for the relative minimum as described above. We see that a relative minimum function value of approximately -1.604 occurs when $x \approx 4.082$.

GRAPHING FUNCTIONS DEFINED PIECEWISE

Section 1.5, Example 5 Graph

$$f(x) = \begin{cases} 4, & \text{for } x \le 0, \\ 4 - x^2, & \text{for } 0 < x \le 2, \\ 2x - 6, & \text{for } x > 2. \end{cases}$$

We will create a multi-statement, user-defined function to graph this function. It is presented here first in block form.

    Func

        If x<=0 Then

        Return 4

      ElseIf x>0 and x<2 Then

        Return 4-x^2

      Else

        Return 2x-6

      EndIf

      EndFunc

When this function is entered on the equation-editor screen, the entire function must be entered on a single line with a colon used to separate statements. That is, we enter Func:If x<=0 Then:Return 4: . . . :EndIf:EndFunc.

Letters must be capitalized and spaces used exactly as shown above. To type a capital letter press ⎡↑⎤ followed by the letter key. To type a single lower case letter, press ⎡alpha⎤ and then the letter key. To turn on the lower case alpha-lock press ⎡2nd⎤ ⎡a-lock⎤. (a-lock is the second operation associated with the ⎡alpha⎤ key.) Press ⎡alpha⎤ to turn off the alpha-lock. The : is the second operation associated with the 4 numeric key. Enter a space by pressing ⎡alpha⎤ and then the ⎡(−)⎤ key.

After the function has been entered, the equation-editor screen will display "y1 = Func."

Now select the Dot graph style. If this is not done, a vertical line that is not part of the graph will appear. Dot style can be selected from the Style menu by highlighting Func on the equation-editor screen and pressing ⎡2nd⎤ ⎡F6⎤ 2 or ⎡2nd⎤ ⎡F6⎤ ⎡▽⎤ ⎡ENTER⎤. Choose and enter window dimensions and then press ⎡◇⎤ ⎡GRAPH⎤ to see the graph of the function. It is shown here in the window $[-5, 5, -3, 6]$.

THE GREATEST INTEGER FUNCTION

The greatest integer function is found in the Catalog and is denoted "int." To find int(1.9) first press $\boxed{\text{CATALOG}}$ $\boxed{\text{I}}$ to go to the first item in the Catalog that begins with I. (I is the purple alphabetic operation associated with the 9 numeric key.) Note that it is not necessary to press $\boxed{\text{alpha}}$ before $\boxed{\text{I}}$ when the Catalog is displayed. Now move the triangular selection cursor beside "int(" and press $\boxed{\text{ENTER}}$ to copy "int(" to the entry line of the home screen. Then press 1 $\boxed{.}$ 9 $\boxed{)}$ $\boxed{\text{ENTER}}$. We can also type "int" directly on the entry line of the home screen. Press $\boxed{\text{2nd}}$ $\boxed{\text{a-lock}}$ $\boxed{\text{I}}$ $\boxed{\text{N}}$ $\boxed{\text{T}}$ $\boxed{\text{alpha}}$ to do this. Then finish entering int(1.9) by pressing $\boxed{(}$ 1 $\boxed{.}$ 9 $\boxed{)}$ $\boxed{\text{ENTER}}$.

We can also graph the greatest integer function.

Section 1.5, Example 7 Graph $f(x) = \text{int}(x)$.

First enter the equation on the equation-editor screen. To do this, clear any existing entries as described on page 67 of this manual and then position the cursor beside y1 =. Copy "int(" from the catalog as described above. Then press $\boxed{\text{X}}$ $\boxed{)}$ $\boxed{\text{ENTER}}$.

Now we set the calculator in the Dot graph style. In the Line graph style, which we have used previously, the calculator will connect adjacent points, or pixels, with line segments and produce a graph that looks like stair-steps rather than a set of horizontal line segments. To select the Dot graph style, highlight the equation and press $\boxed{\text{2nd}}$ $\boxed{\text{F6}}$ 2 or $\boxed{\text{2nd}}$ $\boxed{\text{F6}}$ $\boxed{\triangledown}$ $\boxed{\text{ENTER}}$.

Now choose a window and press ◇ GRAPH . The window $[-6, 6, -6, 6]$ is shown here.

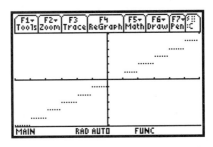

THE COMPOSITION OF FUNCTIONS

We can evaluate composite functions on a graphing calculator.

Section 1.6, Example 4 (b) Given that $f(x) = 2x - 5$ and $g(x) = x^2 - 3x + 8$, find $(f \circ g)(7)$ and $(g \circ f)(7)$.

On the equation-editor screen enter $y_1 = 2x - 5$ and $y_2 = x^2 - 3x + 8$. Then $(f \circ g)(7) = (y_1 \circ y_2)(7)$, or $y_1(y_2(7))$ and $(g \circ f)(7) = (y_2 \circ y_1)(7)$, or $y_2(y_1(7))$. To find these function values press HOME or 2nd QUIT to go to the home screen. Then enter $y_1(y_2(7))$ by pressing Y 1 (Y 2 (7)) ENTER . Now enter $y_2(y_1(7))$ by pressing Y 2 (Y 1 (7)) ENTER .

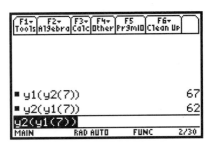

Chapter 2
Functions, Equations and Inequalities

THE INTERSECT METHOD

We can use the Intersection feature from the Math menu on the GRAPH screen to solve equations. We call this the **Intersect method**.

Section 2.1, Example 1 Solve $6x - 4 = 1$.

Press $\boxed{\diamond}$ $\boxed{Y =}$ to go to the equation-editor screen. Clear any existing entries and then enter $y_1 = 6x - 4$ and $y_2 = 1$. The solution of the original equation is the first coordinate of the point of intersection of the graphs of y_1 and y_2. Graph the equations in a window that displays the point of intersection. The standard window is a good choice.

We will use the Intersection feature to find the coordinates of the point of intersection. Press $\boxed{F5}$ 5 to select Intersection from the Math menu on the Graph screen. The query "1st curve?" appears at the bottom of the screen. The blinking cursor is positioned on the graph of y_1. This is indicated by the 1 in the upper right-hand corner of the screen. Press $\boxed{ENTER}$ to indicate that this is the first curve involved in the intersection. Next the query "2nd curve?" appears at the bottom of the screen. The blinking cursor is now positioned on the graph of y_2 and the notation 2 should appear in the top right-hand corner of the screen. Press $\boxed{ENTER}$ to indicate that this is the second curve. We identify the curves for the calculator since we could have more than two graphs on the screen at once. After we identify the second curve, the query "Lower bound?" appears at the bottom of the screen. Use the left or right arrow key to move the blinking cursor to a point to the left of the point of intersection of the lines or type an x-value less than the x-coordinate of the point of intersection. Then press $\boxed{ENTER}$. Next the query "Upper bound?" appears. Move the cursor to a point to the right of the point of intersection or type an x-value greater than the x-value of the point of intersection and press $\boxed{ENTER}$. We give a lower and an upper bound since some pairs of curves have more than one point of intersection. Now the coordinates of the point of intersection appear at the bottom of the screen. We see that the graphs intersect at the point $(0.83333333, 1)$.

THE ZERO METHOD

When an equation is expressed in the form $f(x) = 0$, it can be solved using the Zero feature from the Math menu on the Graph screen.

Section 2.1, Example 8   Find the zero of $f(x) = 5x - 9$.

On the equation-editor screen, clear any existing entries and then enter $y_1 = 5x - 9$. Now graph the function in a viewing window that shows the x-intercept clearly. The standard window is a good choice.

Press $\boxed{\text{F5}}$ 2 to select the Zero feature from the Math menu. We are prompted to select a lower bound. This means that we must choose an x-value that is to the left of the x-intercept. This can be done by using the left- and right-arrow keys to move to a point on the curve to the left of the x-intercept or by keying in a value less than the x-coordinate of the intercept.

Once this is done press $\boxed{\text{ENTER}}$. Now we are prompted to select an upper bound that is to the right of the x-intercept. Again, this can be done by using the arrow keys to move to a point on the curve to the right of the x-intercept or by keying in a value greater than the x-coordinate of the intercept.

Press $\boxed{\text{ENTER}}$ again. We see that $y = 0$ when $x = 1.8$, so 1.8 is the zero of the function f.

OPERATIONS WITH COMPLEX NUMBERS

Operations with complex numbers can be performed on the TI-89. First set the calculator in the rectangular mode by pressing $\boxed{\text{MODE}}$, highlighting the entry for Complex Format, and then pressing $\boxed{\triangleright}$ 2 $\boxed{\text{ENTER}}$ or $\boxed{\triangleright}$ $\boxed{\triangledown}$ $\boxed{\text{ENTER}}$ $\boxed{\text{ENTER}}$.

Section 2.2, Example 2

(a) Add: $(8 + 6i) + (3 + 2i)$.

To find this sum go to the home screen and press 8 $\boxed{+}$ 6 $\boxed{\text{2nd}}$ $\boxed{i}$ $\boxed{+}$ 3 $\boxed{+}$ 2 $\boxed{\text{2nd}}$ $\boxed{i}$ $\boxed{\text{ENTER}}$. (The number i is the second operation associated with the $\boxed{\text{CATALOG}}$ key.) Note that it is not necessary to include parentheses when we are adding.

(b) Subtract: $(4 + 5i) - (6 - 3i)$.

Press 4 $\boxed{+}$ 5 $\boxed{\text{2nd}}$ $\boxed{i}$ $\boxed{-}$ $\boxed{(}$ 6 $\boxed{-}$ 3 $\boxed{\text{2nd}}$ $\boxed{i}$ $\boxed{)}$ $\boxed{\text{ENTER}}$. Note that the parentheses must be included as shown so that the entire number $6 - 3i$ is subtracted.

Section 2.2, Example 3

(a) Multiply: $\sqrt{-16} \cdot \sqrt{-25}$.

Press $\boxed{\text{2nd}}$ $\boxed{\sqrt{}}$ $\boxed{(-)}$ 1 6 $\boxed{)}$ $\boxed{\text{2nd}}$ $\boxed{\sqrt{}}$ $\boxed{(-)}$ 2 5 $\boxed{)}$ $\boxed{\text{ENTER}}$.

(b) Multiply: $(1 + 2i)(1 + 3i)$.

Press $\boxed{(}$ 1 $\boxed{+}$ 2 $\boxed{\text{2nd}}$ $\boxed{i}$ $\boxed{)}$ $\boxed{(}$ 1 $\boxed{+}$ 3 $\boxed{\text{2nd}}$ $\boxed{i}$ $\boxed{)}$ $\boxed{\text{ENTER}}$.

(c) Multiply: $(3 - 7i)^2$.

Press $\boxed{(}$ 3 $\boxed{-}$ 7 $\boxed{\text{2nd}}$ $\boxed{i}$ $\boxed{)}$ $\boxed{\wedge}$ 2 $\boxed{\text{ENTER}}$.

Chapter 3
Polynomial and Rational Functions

POLYNOMIAL MODELS

We can fit second-degree, or quadratic, functions, third-degree, or cubic, functions and fourth-degree, or quartic, functions to data on a graphing calculator.

Section 3.1, Example 9(a) The table below shows the number of farms, in millions, for several years beginning with 1900. Model the data with both cubic and quartic functions. Let the first coordinate of each data point be the number of years after 1900. Then using R^2, the coefficient of determination, decide which function is the better fit. The closer R^2 is to 1, the better the fit.

Years after 1900	Number of Farms, in millions
0, 1900	5.740
10, 1910	6.366
20, 1920	6.454
30, 1930	6.295
40, 1940	6.102
50, 1950	5.388
60, 1960	3.962
70, 1970	2.954
80, 1980	2.440
90, 1990	2.146
95, 1995	2.196
100, 2000	2.172
102, 2002	2.158

Enter the data in lists as described on page 74 of this manual. To model the data with a cubic function select cubic regression from the Calculate menu by pressing $\boxed{F5}$ $\boxed{\triangleright}$ 3 $\boxed{\text{ENTER}}$. Indicate that the data in c1 and c2 will be used for x and y, respectively, and that the equation should be copied to the equation-editor screen as y_1 as described on page 76 of this manual. The graphing calculator displays the coefficients of a cubic function $y = ax^3 + bx^2 + cx + d$ along with the value of R^2.

After the data are entered, we can also find a quartic model by pressing [F5] [▷] [alpha] [A] [ENTER] to select quartic regression from the Calculate menu. The calculator displays the coefficients of a quartic function $y = ax^4 + bx^3 + cx^2 + dx + e$ along with the value for R^2.

Since the R^2-value for the quartic function is closer to 1 than that for the cubic function, the quartic function is the better fit.

A scatterplot of the data can be graphed as described on page 75 of this manual. The function has been copied to the Y = screen, so it can be graphed along with the scatterplot. It can also be evaluated using one of the methods on pages 71 - 73.

When appropriate, a quadratic model for data can be found by selecting quadratic regression, item 9, from the Clculate menu after the data are entered. This will be demonstrated in Chapter 5.

Chapter 4
Exponential and Logarithmic Functions

GRAPHING AN INVERSE FUNCTION

The DrawInv operation can be used to graph a function and its inverse on the same screen. A formula for the inverse function need not be found in order to do this. The calculator must be set in Func mode when this operation is used.

Section 4.1, Example 7 Graph $f(x) = 2x - 3$ and $f^{-1}(x)$ using the same set of axes.

Enter $y_1 = 2x - 3$, clear all other functions on the Y = screen, and choose a viewing window. Press $\boxed{\text{HOME}}$ to go to the home screen. Then press $\boxed{\text{CATALOG}}$ $\boxed{\text{D}}$ to go to the first item in the Catalog that begins with D. Use the $\boxed{\triangledown}$ key to scroll down to DrawInv and then press $\boxed{\text{ENTER}}$ to copy DrawInv to the entry line of the home screen. Follow these keystrokes with $\boxed{\text{Y}}$ $\boxed{1}$ $\boxed{(}$ $\boxed{\text{X}}$ $\boxed{)}$ to select function y_1. Press $\boxed{\text{ENTER}}$ to see the graph of the function and its inverse. The graphs are shown here in the standard window.

EVALUATING e^x, **Log** x, **and Ln** x

We can use the calculator's scientific keys to evaluate e^x, $\log x$, and $\ln x$ for specific values of x.

Section 4.2, Example 5 (a), (b) Find the value of e^3 and $e^{-0.23}$. Round to four decimal places.

First select Approximate mode by pressing $\boxed{\text{MODE}}$ $\boxed{\text{F2}}$ $\boxed{\triangledown}$ $\boxed{\triangledown}$ $\boxed{\triangleright}$ $\boxed{3}$ $\boxed{\text{ENTER}}$. Then to find e^3 press $\boxed{\diamond}$ $\boxed{e^x}$ $\boxed{3}$ $\boxed{)}$ $\boxed{\text{ENTER}}$. (e^x is the green $\diamond$ operation associated with the $\boxed{\text{X}}$ key.) The calculator returns 20.0855369232. Thus, $e^3 \approx 20.0855$. To find $e^{-0.23}$ press $\boxed{\diamond}$ $\boxed{e^x}$ $\boxed{(-)}$ $\boxed{\cdot}$ $\boxed{2}$ $\boxed{3}$ $\boxed{)}$ $\boxed{\text{ENTER}}$. The calculator returns .794533602503, so $e^{-0.23} \approx 0.7945$.

Section 4.3, Example 5 Find the values of log 645,778, log 0.0000239, and log (-3). Round to four decimal places.

The calculator must be in Approximate mode as described in Example 6 above. To find log 645,778 press $\boxed{\text{2nd}}$ $\boxed{\text{a-lock}}$ $\boxed{\text{L}}$ $\boxed{\text{O}}$ $\boxed{\text{G}}$ $\boxed{\text{alpha}}$ $\boxed{(}$ 6 4 5 7 7 8 $\boxed{)}$ $\boxed{\text{ENTER}}$ and read 5.81008324563. Thus, log $645,778 \approx 5.8101$. The operation "log(" can also be selected from the Catalog. To find log 0.0000239 press $\boxed{\text{2nd}}$ $\boxed{\text{a-lock}}$ $\boxed{\text{L}}$ $\boxed{\text{O}}$ $\boxed{\text{G}}$ $\boxed{\text{alpha}}$ $\boxed{(}$ $\boxed{\cdot}$ 0 0 0 0 2 3 9 $\boxed{)}$ $\boxed{\text{ENTER}}$. The calculator returns -4.62160209905, so log $0.0000239 \approx -4.6216$. The previous entry, log $645,778$, can also be edited to find log 0.0000239. When the TI-89 has Real selected for the Complex Format mode the keystrokes $\boxed{\text{2nd}}$

a-lock ⎴L⎴ ⎴O⎴ ⎴G⎴ ⎴alpha⎴ ⎴(⎴ ⎴(−)⎴ ⎴3⎴ ⎴)⎴ ⎴ENTER⎴ produce the error message "Non-real result" indicating that the result of this calculation is not a real number. We could also have entered log(−3) by editing the previous entry.

Section 4.3, Example 6 (a), (b), (c) Find the values of ln 645,778, ln 0.0000239, and ln (−5). Round to four decimal places.

The calculator must be in Approximate mode as described in Example 6 above. To find ln 645,778 and ln 0.0000239 repeat the keystrokes used above to find log 645,778 and log 0.0000239 but press ⎴2nd⎴ ⎴LN⎴ rather than ⎴2nd⎴ ⎴a-lock⎴ ⎴L⎴ ⎴O⎴ ⎴G⎴ ⎴alpha⎴ ⎴(⎴. (LN is the second operation associated with the ⎴X⎴ key.) We find that ln 645,778 ≈ 13.3782 and ln 0.0000239 ≈ −10.6416. When the TI-89 has Real selected for the Complex Format mode the keystrokes ⎴2nd⎴ ⎴LN⎴ ⎴(−)⎴ ⎴5⎴ ⎴)⎴ ⎴ENTER⎴ produce the error message "Non-real result" indicating that the result of this calculation is not a real number.

USING THE CHANGE OF BASE FORMULA

To find a logarithm with a base other than 10 or e we use the change-of-base formula, $\log_b M = \dfrac{\log_a M}{\log_a b}$, where a and b are any logarithmic bases and M is any positive number.

Section 4.3, Example 7 Find $\log_5 8$ using common logarithms.

The calculator must be in Approximate mode as described in Example 6 above. We let $a = 10$, $b = 5$, and $M = 8$ and substitute in the change-of-base formula. Press ⎴2nd⎴ ⎴a-lock⎴ ⎴L⎴ ⎴O⎴ ⎴G⎴ ⎴alpha⎴ ⎴(⎴ ⎴8⎴ ⎴)⎴ ⎴÷⎴ ⎴2nd⎴ ⎴a-lock⎴ ⎴L⎴ ⎴O⎴ ⎴G⎴ ⎴alpha⎴ ⎴(⎴ ⎴5⎴ ⎴)⎴ ⎴ENTER⎴. The result is about 1.2920. We could have let $a = e$ and used natural logarithms to find $\log_5 8$ as well.

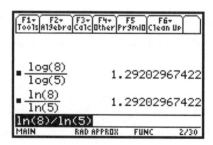

Section 4.3, Example 9 Graph $y = \log_5 x$.

To use a TI-89 we must first change the base to e or 10. Here we use e. Let $a = e$, $b = 5$, and $M = x$ and substitute in the change-of-base formula. Enter $y_1 = \dfrac{\ln x}{\ln 5}$ on the Y = screen, select a window, and press ⎴◇⎴ ⎴GRAPH⎴.

EXPONENTIAL AND LOGARITHMIC REGRESSION

In addition to the types of polynomial regression discussed earlier, exponential and logarithmic functions can be fit to data. The operations of entering data, making scatterplots, and graphing and evaluating these functions are the same as for linear regression functions. So are the procedures for copying a regression equation to the Y = screen, graphing it, and using it to find function values.

Section 4.6, Example 6 (a) *Debit Card Volume.* The total debit card volume has increased dramatically in recent years, as shown in the table below. Use a graphing calculator to fit an exponential function to the data.

Year, x	Debit Card Volume, y (in billions)
1997, 0	162.5
1998, 1	238.5
1999, 2	322.6
2000, 3	419.1
2001, 4	571.8

Enter the data in the Data/Matrix editor as described on page 74 of this manual. Then select exponential regression from the Calculate menu by pressing F5 ▷ 4 ENTER. Indicate that the data in c1 and c2 will be used for x and y, respectively, and that the equation should be copied to the equation-editor screen as y_1 as described on page 76 of this manual. The calculator displays the coefficient a and the base b for the exponential function $y = a \cdot b^x$.

A scatterplot of the data can be graphed as described on page 75 of this manual. The function has been copied to the Y = screen, so it can be graphed along with the scatterplot. It can also be evaluated using one of the methods on pages 71 - 73.

Section 4.6, Exercise 28 *Forgetting.* In an art class, students were tested at the end of the course on a final exam. Then they were retested with an equivalent test at subsequent time intervals. Their scores after time x, in months, are given in the following table. Use a graphing calculator to fit a logarithmic function $y = a + b \ln(x)$ to the data.

Time, x (in months)	Score, y
1	84.9%
2	84.6%
3	84.4%
4	84.2%
5	84.1%
6	83.9%

After entering the data in the Data/Matrix editor as described on page 74 of this manual, press $\boxed{\text{F5}}$ $\boxed{\triangleright}$ 6 $\boxed{\text{ENTER}}$ to go to the Calculate menu and select LnReg. Indicate that the data in c1 and c2 will be used for x and y, respectively, and that the equation should be copied to the equation-editor screen as y_1 as described on page 76 of this manual. The values of a and b for the logarithmic function $y = a + b\ln(x)$ are displayed.

LOGISTIC REGRESSION

A logistic function can be fit to data using the TI-89.

Section 4.6, Exercise 32 *Effect of Advertising.* A company introduces a new software product on a trial run in a city. They advertised the product on television and found the following data relating the percent P of people who bought the product after x ads were run. Use a graphing calculator to fit a logistic function to the data.

Number of Ads, x	Percent Who Bought, P
0	0.2
10	0.7
20	2.7
30	9.2
40	27
50	57.6
60	83.3
70	94.8
80	98.5
90	99.6

After entering the data in the Data/Matrix editor as described on page 74 of this manual, press $\boxed{\text{F5}}$ $\boxed{\triangleright}$ $\boxed{\text{alpha}}$ $\boxed{\text{C}}$ to go to the Calculate menu and select Logistic. Indicate that the data in c1 and c2 will be used for x and y, respectively, and that the equation should be copied to the equation-editor screen as y_1 as described on page 76 of this manual. The values of a, b, and c for the logistic function $y = a/(1 + be^{cx}) + d$ are displayed.

Chapter 5
Systems of Equations and Matrices

QUADRATIC REGRESSION

Quadratic functions can be fit to data using the quadratic regression operation. The operations of entering data, making scatterplots, and graphing and evaluating quadratic regression functions are the same as for linear regression functions.

Section 5.2, Exercise 35 *Morning Newspapers* The number of morning newspapers in the United States in various years is shown in the following table.

Year	Number of Morning Newspapers
1920	437
1940	380
1960	312
1980	387
1990	559
2001	776

(a) Use a graphing calculator to fit a quadratic function to the data, where x is the number of years after 1920.

(b) Use the function found in part (a) to estimate the number of morning newspapers in 2006 and 2009.

(a) Clear any existing entries on the equation-editor screen. Then enter the data in the Data/Matrix editor as described on page 74 of this manual. To fit a quadratic function to the data, press $\boxed{\text{APPS}}$ 6 $\boxed{\text{ENTER}}$ $\boxed{\text{F5}}$ to view the Calculate menu. Then select QuadReg by pressing $\boxed{\triangleright}$ 9. Then press $\boxed{\triangledown}$ $\boxed{\text{alpha}}$ $\boxed{\text{C}}$ 1 $\boxed{\triangledown}$ $\boxed{\text{alpha}}$ $\boxed{\text{C}}$ 2 to indicate that the data in c1 and c2 will be used for x and y, respectively. Press $\boxed{\triangledown}$ $\boxed{\triangleright}$ $\boxed{\triangledown}$ $\boxed{\text{ENTER}}$ to indicate that the regression equation should be copied to the equation-editor screen as y_1. Finally press $\boxed{\text{ENTER}}$ again to see the STAT VARS screen which displays the coefficients $a, b,$ and c of the regression equation $y = ax^2 + bx + c$. Note that at least three data points are required for quadratic regression.

(b) To estimate the number of morning newspapers in 2006 and 2009, we evaluate the regression function for 86 and 89, respectively. We can use any of the methods for evaluating a function found on pages 71 - 73 of this manual. Here we show a table set in Ask mode.

MATRICES AND ROW-EQUIVALENT OPERATIONS

Matrices with up to 999 rows and 99 columns can be entered on a TI-89. Row-equivalent operations can be performed on matrices on the calculator.

Section 5.3, Example 1 Solve the following system:

$$2x - y + 4z = -3,$$
$$x - 2y - 10z = -6,$$
$$3x + 4z = 7.$$

First we enter the augmented matrix

$$\begin{bmatrix} 2 & -1 & 4 & -3 \\ 1 & -2 & -10 & -6 \\ 3 & 0 & 4 & 7 \end{bmatrix}$$

in the Data/Matrix editor. We will call the Matrix A. Press $\boxed{\text{APPS}}$ 6 3 $\boxed{\triangleright}$ 2 $\boxed{\triangledown}$ $\boxed{\triangledown}$ $\boxed{\text{alpha}}$ $\boxed{\text{A}}$ $\boxed{\triangledown}$ 3 $\boxed{\triangledown}$ 4 $\boxed{\text{ENTER}}$ $\boxed{\text{ENTER}}$ to go to the Data/Matrix editor and set up a matrix named A with 3 rows and 4 columns. If a matrix named A has previously been saved in your calculator, an error message will be displayed. If this happens, you can select a different name for the matrix we are about to enter or you can delete the current matrix A and then enter the new matrix as A. To delete a matrix press $\boxed{\text{2nd}}$ $\boxed{\text{VAR-LINK}}$, use $\boxed{\triangledown}$ to highlight the name of the matrix being deleted, and then press $\boxed{\text{F1}}$ 1 $\boxed{\text{ENTER}}$ or $\boxed{\text{F1}}$ $\boxed{\text{ENTER}}$ $\boxed{\text{ENTER}}$. (VAR-LINK is the second operation associated with the $\boxed{-}$ key.)

Enter the elements of the first row of the matrix by pressing 2 $\boxed{\text{ENTER}}$ $\boxed{(-)}$ 1 $\boxed{\text{ENTER}}$ 4 $\boxed{\text{ENTER}}$ $\boxed{(-)}$ 3 $\boxed{\text{ENTER}}$. The cursor moves to the element in the second row and first column of the matrix. Enter the elements of the second and third rows of the augmented matrix by typing each in turn followed by $\boxed{\text{ENTER}}$ as above. Note that the screen only displays three columns of the matrix. The arrow keys can be used to move the cursor to any element at any time.

```
┌─────┬──────────┬─────┬──────┬──────┬────────┐
│ F1▾ │    F2    │ F3  │      │      │ F6▾│F7 │
│Tools│Plot Setup│Cell │      │      │Util│Stat│
├─────┴──────────┴─────┴──────┴──────┴────┴────┤
│MAT                                           │
│3x4                                           │
│      ┌─────────┬─────────┬─────────┐         │
│      │c1       │c2       │c3       │         │
│1     │2.       │-1.      │4.       │         │
│2     │1.       │-2.      │-10.     │         │
│3     │3.       │0.       │4.       │         │
│4     │         │         │         │         │
│      └─────────┴─────────┴─────────┘         │
│r1c1=2.                                       │
├──────────────────────────────────────────────┤
│MAIN        RAD APPROX    FUNC                │
└──────────────────────────────────────────────┘
```

Matrix operations are performed on the home screen and are found on the Math Matrix menu. Press [HOME] or [2nd] [QUIT] to leave the matrix editor and go to this screen. Access the Math Matrix menu by pressing [2nd] [MATH] 4. (MATH is the second operation associated with the 5 numeric key.)

Press [alpha] [J] to see the four row-equivalent operations: rowSwap, rowAdd, nRow, and nRowAdd. These operations interchange two rows of a matrix, add two rows, multiply a row by a number, and multiply a row by a number and add it to a second row, respectively.

We will use the calculator to perform the row-equivalent operations that were done algebraically in the text. First, to interchange row 1 and row 2 of matrix **A**, with the MATH Matrix menu displayed, press 1 to select rowSwap. Then press [alpha] [A] to select **A**. Follow this with a comma and the rows to be interchanged, [,] 1 [,] 2 [)] [ENTER].

```
┌─────┬──────┬─────┬─────┬──────┬─────────┐
│ F1▾ │ F2▾  │ F3▾ │ F4▾ │ F5   │  F6▾    │
│Tools│Algebra│Calc│Other│PrgmIO│Clean Up │
├─────┴──────┴─────┴─────┴──────┴─────────┤
│                                         │
│                                         │
│ ■ rowSwap(a,1,2)                        │
│        ┌                      ┐         │
│        │1.    -2.    -10.   -6.│        │
│        │2.    -1.    4.     -3.│        │
│        │3.    0.     4.     7. │        │
│        └                      ┘         │
│rowSwap(a,1,2)                           │
├─────────────────────────────────────────┤
│MAIN        RAD APPROX    FUNC           │
└─────────────────────────────────────────┘
```

The calculator will not store the matrix produced using a row-equivalent operation, so when several operations are to be performed in succession it is helpful to store the result of each operation as it is produced. For example, to store the matrix resulting from interchanging the first and second rows of **A** as matrix **B** press [STO▷] [alpha] [B] [ENTER] immediately after interchanging the rows. This can also be done before [ENTER] is pressed at the end of the rowSwap.

Next we multiply the first row of **B** by -2, add it to the second row and store the result as **B** again by pressing [2nd] [MATH] 4 [alpha] [J] 4 [(−)] 2 [,] [alpha] [B] [,] 1 [,] 2 [)] [STO▷] [alpha] [B] [ENTER]. These keystrokes select mRowAdd(from the MATH Matrix menu; then they specify that the value of the multiplier is -2, the matrix being operated on is **B**, and that a multiple of row 1 is being added to row 2; finally they store the result as **B**.

To multiply row 1 by -3, add it to row 3, and store the result as **B** press [2nd] [MATH] 4 [alpha] [J] 4 [(−)] 3 [,] [alpha] [B] [,] 1 [,] 3 [)] [STO▷] [alpha] [B] [ENTER].

Now multiply the second row by 1/3 and store the result as **B** again. Press 2nd MATH 4 alpha J 3 1 (÷) 3 , alpha B , 2) STO▷ alpha B ENTER . These keystrokes select mRow(from the MATH Matrix menu; then they specify that the value of the multiplier is 1/3, the matrix being operated on is **B**, and row 2 is being multiplied; finally they store the result as **B**.

Multiply the second row by -6 and add it to the third row using mRowAdd(. Press 2nd MATH 4 ALPHA J 4 (−) 6 , alpha B , 2 , 3) STO▷ alpha B ENTER . The entry in the third row, second column is 1E−13. This is an approximation of 0 that occurs because of the manner in which the calculator performs calculations and should be treated as 0. In fact, it would be a good idea to return to the Data/Matrix editor at this point to replace this entry of **B** with 0. Press APPS 6 2 ▷ 2 ▽ ▽ ▷ . Then highlight **B** and press ENTER ENTER . Now highlight the entry in the third, row, second column and press 0 ENTER . Press HOME alpha B ENTER to see the edited version of **B**.

Finally, multiply the third row by $-1/14$ by pressing 2nd MATH 4 ALPHA J 3 (−) 1 (÷) 1 4 , alpha B , 3) ENTER .

Write the system of equations that corresponds to the final matrix. Then use back-substitution to solve for x, y, and z as illustrated in the text.

Section 5.3, Example 3 Instead of stopping with row-echelon form as we did in Example 1 above, we can continue to apply row-equivalent operations until the matrix is in reduced row-echelon form. Reduced row-echelon form of a matrix can be found directly by using the rref(operation from the MATH Matrix menu. For example, to find reduced row-echelon form for matrix **A** in Example 1 above, after entering **A** and leaving the Data/Matrix screen [2nd] [MATH] 4 4 [alpha] [A] [)] [ENTER]. We can read the solution of the system of equations, $(3, 7, -0.5)$ directly from the resulting matrix.

MATRIX OPERATIONS

We can use a graphing calculator to add and subtract matrices, to multiply a matrix by a scalar, and to multiply matrices.

Section 5.4, Example 1 (a) Find $\mathbf{A} + \mathbf{B}$ for

a) $\mathbf{A} = \begin{bmatrix} -5 & 0 \\ 4 & \frac{1}{2} \end{bmatrix}$, $\mathbf{B} = \begin{bmatrix} 6 & -3 \\ 2 & 3 \end{bmatrix}$.

Enter **A** and **B** in the Data/Matrix editor as described earlier in this chapter of the Graphing Calculator Manual. If you used the matrix names **A** and **B** above in Example 1 from Section 5.3, it will be necessary to delete those matrices before matrices **A** and **B** from this example can be entered. See page 94 of this manual for the procedure. Press [HOME] or [2nd] [QUIT] to leave the Data/Matrix screen. Then press [alpha] [A] [+] [alpha] [B] [ENTER] to display the sum.

Section 5.4, Example 2 Find $\mathbf{C} - \mathbf{D}$ for each of the following.

a) $\mathbf{C} = \begin{bmatrix} 1 & 2 \\ -2 & 0 \\ -3 & -1 \end{bmatrix}$, $\mathbf{D} = \begin{bmatrix} 1 & -1 \\ 1 & 3 \\ 2 & 3 \end{bmatrix}$ b) $\mathbf{C} = \begin{bmatrix} 5 & -6 \\ -3 & 4 \end{bmatrix}$, $\mathbf{D} = \begin{bmatrix} -4 \\ 1 \end{bmatrix}$

a) Enter $\mathbf{C}$ and $\mathbf{D}$ in the Data/Matrix editor. Press HOME or 2nd QUIT to leave this screen. Then press alpha C − alpha D ENTER to display the difference.

b) Enter $\mathbf{C}$ and $\mathbf{D}$ in the Data/Matrix editor after deleting $\mathbf{C}$ and $\mathbf{D}$ from part(a) above if necessary. (See page 94 of this manual for the procedure.) Press HOME or 2nd QUIT to leave the Data/Matrix screen. Then press alpha C − alpha D ENTER. The calculator returns the error message "Dimension mismatch," indicating that this subtraction is not possible. This is the case because the matrices have different orders.

Section 5.4, Example 4 Find $3\mathbf{A}$ and $(-1)\mathbf{A}$, for $\mathbf{A} = \begin{bmatrix} -3 & 0 \\ 4 & 5 \end{bmatrix}$

Enter $\mathbf{A}$ in the Data/Matrix editor after deleting the matrix that was previously entered as $\mathbf{A}$ if necessary. (See page 94 of this manual for the procedure.) Press HOME or 2nd QUIT to leave the Data/Matrix screen. Then to find $3\mathbf{A}$ press 3 alpha A ENTER and to find $(-1)\mathbf{A}$ press (−) 1 alpha A ENTER. Note that $(-1)\mathbf{A}$ is the opposite, or additive inverse, of $\mathbf{A}$ and can also be found by pressing (−) alpha A ENTER.

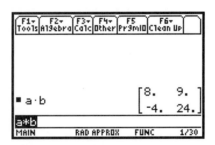

Section 5.4, Example 6 (a), (d) For

$$\mathbf{A} = \begin{bmatrix} 3 & 1 & -1 \\ 2 & 0 & 3 \end{bmatrix}, \mathbf{B} = \begin{bmatrix} 1 & 6 \\ 3 & -5 \\ -2 & 4 \end{bmatrix}, \text{ and } \mathbf{C} = \begin{bmatrix} 4 & -6 \\ 1 & 2 \end{bmatrix}$$

find each of the following.

a) **AB**   d) **AC**

First enter **A**, **B**, and **C** in the Data/Matrix editor after first deleting any matrices that were previously entered with those names. (See page 94 of this manual for the procedure.) Then press $\boxed{\text{HOME}}$ or $\boxed{\text{2nd}}$ $\boxed{\text{QUIT}}$ to leave this screen.

a) To find **AB** press $\boxed{\text{alpha}}$ $\boxed{\text{A}}$ $\boxed{\times}$ $\boxed{\text{alpha}}$ $\boxed{\text{B}}$ $\boxed{\text{ENTER}}$. Note that the multiplication symbol must be used so that the calculator can differentiate this multiplication from a variable named *ab*.

d) To find **AC** press $\boxed{\text{alpha}}$ $\boxed{\text{A}}$ $\boxed{\times}$ $\boxed{\text{alpha}}$ $\boxed{\text{C}}$ $\boxed{\text{ENTER}}$. The calculator returns the error message "Dimension," indicating that this multiplication is not possible. This is the case because the number of columns in **A** is not the same as the number of rows in **C**. Thus, the matrices cannot be multiplied in this order.

FINDING THE INVERSE OF A MATRIX

The inverse of a matrix can be found quickly on a graphing calculator.

Section 5.5, Example 3 Find $\mathbf{A}^{-1}$, where

$$\mathbf{A} = \begin{bmatrix} -2 & 3 \\ -3 & 4 \end{bmatrix}.$$

Enter **A** in the Data/Matrix editor after first deleting the matrix that was previously entered as **A** if necessary. (See page 94 of this manual for the procedure.) Then press $\boxed{\text{HOME}}$ or $\boxed{\text{2nd}}$ $\boxed{\text{QUIT}}$ to leave the Data Matrix screen. Now press $\boxed{\text{alpha}}$ $\boxed{\text{A}}$ $\boxed{\wedge}$ $\boxed{(-)}$ 1 $\boxed{\text{ENTER}}$.

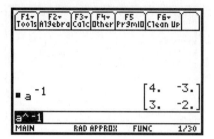

Section 5.5, Exercise 7 Find $\mathbf{A}^{-1}$, where

$$\mathbf{A} = \begin{bmatrix} 6 & 9 \\ 4 & 6 \end{bmatrix}.$$

Enter $\mathbf{A}$ in the Data/Matrix editor after first deleting the matrix that was previously entered as $\mathbf{A}$ if necessary. (See page 94 of this manual for the procedure.) Then press $\boxed{\text{HOME}}$ or $\boxed{\text{2nd}}$ $\boxed{\text{QUIT}}$ to leave the Data Matrix screen. Now press $\boxed{\text{alpha}}$ $\boxed{\text{A}}$ $\boxed{\wedge}$ $\boxed{(-)}$ 1 $\boxed{\text{ENTER}}$. The calculator returns the message "Singular matrix," indicating that $\mathbf{A}^{-1}$ does not exist.

MATRIX SOLUTIONS OF SYSTEMS OF EQUATIONS

We can write a system of n linear equations in n variables as a matrix equation $\mathbf{AX} = \mathbf{B}$. If $\mathbf{A}$ has an inverse the solution of the system of equations is given by $\mathbf{X} = \mathbf{A}^{-1}\mathbf{B}$.

Section 5.5, Example 5 Use an inverse matrix to solve the following system of equations:

$$-2x + 3y = 4,$$
$$-3x + 4y = 5.$$

Enter $\mathbf{A} = \begin{bmatrix} -2 & 3 \\ -3 & 4 \end{bmatrix}$ and $\mathbf{B} = \begin{bmatrix} 4 \\ 5 \end{bmatrix}$ in the Data/Matrix editor after first deleting any matrices that were previously entered with those names. (See page 94 of this manual for the procedure.) Then press $\boxed{\text{HOME}}$ or $\boxed{\text{2nd}}$ $\boxed{\text{QUIT}}$ to leave the Data/Matrix screen. Press $\boxed{\text{alpha}}$ $\boxed{\text{A}}$ $\boxed{\wedge}$ $\boxed{(-)}$ 1 $\boxed{\times}$ $\boxed{\text{alpha}}$ $\boxed{\text{B}}$ $\boxed{\text{ENTER}}$. The result is the 2 x 1 matrix $\begin{bmatrix} 1 \\ 2 \end{bmatrix}$, so the solution is $(1, 2)$.

DETERMINANTS AND CRAMER'S RULE

We can evaluate determinants on a graphing calculator and use Cramer's rule to solve systems of equations.

Section 5.6, Example 5 Use a graphing calculator to evaluate $|\mathbf{A}|$.

$$\mathbf{A} = \begin{bmatrix} 1 & 6 & -1 \\ -3 & -5 & 3 \\ 0 & 4 & 2 \end{bmatrix}$$

Enter $\mathbf{A}$ in the Data/Matrix editor and then press $\boxed{\text{HOME}}$ or $\boxed{\text{2nd}}$ $\boxed{\text{QUIT}}$ to leave this screen. We will select the "det(" operation from the MATH Matrix menu. Press $\boxed{\text{2nd}}$ $\boxed{\text{MATH}}$ 4 2 $\boxed{\text{alpha}}$ $\boxed{\text{A}}$ $\boxed{)}$ $\boxed{\text{ENTER}}$. We see that $|\mathbf{A}| = 26$.

Section 5.6, Example 6 Solve using Cramer's rule:

$$2x + 5y = 7,$$
$$5x - 2y = -3.$$

First we enter the matrices corresponding to D, D_x, and D_y as $\mathbf{A}$, $\mathbf{B}$, and $\mathbf{C}$, respectively. We have

$$\mathbf{A} = \begin{bmatrix} 2 & 5 \\ 5 & -2 \end{bmatrix}, \mathbf{B} = \begin{bmatrix} 7 & 5 \\ -3 & -2 \end{bmatrix}, \text{ and } \mathbf{C} = \begin{bmatrix} 2 & 7 \\ 5 & -3 \end{bmatrix}.$$

Then we use the "det(" operation from the MATH Matrix menu. We have

$$x = \frac{\det(\mathbf{B})}{\det(\mathbf{A})} \text{ and } y = \frac{\det(\mathbf{C})}{\det(\mathbf{A})}.$$

To find x press $\boxed{\text{2nd}}$ $\boxed{\text{MATH}}$ 4 2 $\boxed{\text{alpha}}$ $\boxed{\text{B}}$ $\boxed{)}$ $\boxed{\div}$ $\boxed{\text{2nd}}$ $\boxed{\text{MATH}}$ 4 2 $\boxed{\text{alpha}}$ $\boxed{\text{A}}$ $\boxed{)}$ $\boxed{\text{ENTER}}$. The result is $-\dfrac{1}{29}$. To find y press $\boxed{\text{2nd}}$ $\boxed{\text{MATH}}$ 4 2 $\boxed{\text{alpha}}$ $\boxed{\text{C}}$ $\boxed{)}$ $\boxed{\div}$ $\boxed{\text{2nd}}$ $\boxed{\text{MATH}}$ 4 2 $\boxed{\text{alpha}}$ $\boxed{\text{A}}$ $\boxed{)}$ $\boxed{\text{ENTER}}$. We can also recall the entry for x and edit it, replacing matrix $\mathbf{B}$ with matrix $\mathbf{C}$, to find y. In either case, the result is $\dfrac{41}{29}$. The solution of the system of equations is $\left(-\dfrac{1}{29}, \dfrac{41}{29} \right)$.

GRAPHS OF INEQUALITIES

We can graph linear inequalities on a graphing calculator, shading the region of the solution set. The calculator should be set in Func mode at this point.

Section 5.7, Example 1 Graph: $y < x + 3$.

First we graph the related equation $y = x + 3$. We use the standard window $[-10, 10, -10, 10]$. Since the inequality symbol is $<$ we know that the line $y = x + 3$ is not part of the solution set. In a hand-drawn graph we would use a dashed line to indicate this. After determining that the solution set of the inequality consists of all points below the line, we can use the "shade below" graph style to shade this region. After entering the related equation, highlight it and press $\boxed{\text{2nd}}$ $\boxed{\text{F6}}$ 8. Then press $\boxed{\diamond}$ $\boxed{\text{GRAPH}}$ to display the graph of the inequality. Keep in mind the fact that the line $y = x + 3$ is not included in the solution set.

We can also use the Shade feature from the Catalog to graph this inequality. Copy Shade to the entry line of the home screen by pressing $\boxed{\text{HOME}}$ $\boxed{\text{CATALOG}}$ $\boxed{\text{S}}$, scrolling down to Shade, and pressing $\boxed{\text{ENTER}}$. We can also type Shade directly on the entry line by pressing $\boxed{\text{2nd}}$ $\boxed{\text{a-lock}}$ $\boxed{\text{S}}$ $\boxed{\text{H}}$ $\boxed{\text{A}}$ $\boxed{\text{D}}$ $\boxed{\text{E}}$ $\boxed{\text{alpha}}$.

Now enter a lower function and an upper function. The region between them will be shaded. We want to shade the area between the bottom of the window, $y = -10$, and the line $y = x + 3$ so we enter $\boxed{(-)}$ 1 0 $\boxed{,}$ $\boxed{\text{X}}$ $\boxed{+}$ 3 $\boxed{\text{ENTER}}$. We can also enter $x + 3$ as $y_1(x)$. The result is shown below. Keep in mind that the line $y = x + 3$ is not included in the solution set.

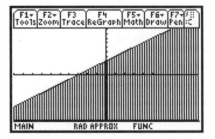

We can also graph a system of inequalities when the solution set lies between the graphs of two functions.

Section 5.7, Exercise 43 Graph:
$$y \le x,$$
$$y \ge 3 - x.$$

First graph the related equations $y_1 = x$ and $y_2 = 3 - x$ and determine that the solution set consists of all the points on or below the graph of $y_1 = x$ and on or above the graph of $y_2 = 3 - x$. We can graph the system of inequalities by shading the solution set of each inequality in the system with a different pattern. When the "shade above" or "shade below" graph style options are selected the TI-89 rotates through four shading patterns. These patterns repeat if more than four functions are graphed. The region where the shaded areas overlap is the solution set of the inequality. Shade below $y_1 = x$ by highlighting the equation and then pressing $\boxed{\text{2nd}}$ $\boxed{\text{F6}}$ 8. Shade above $y_2 = 3 - x$ by highlighting the equation and pressing $\boxed{\text{2nd}}$ $\boxed{\text{F6}}$ 7. Then press $\boxed{\diamond}$ $\boxed{\text{GRAPH}}$.

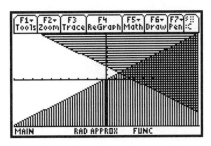

We can also use the Shade feature from the Catalog to graph this system of inequalities. Copy Shade to the entry line of the home screen or type it directly as described in Example 1 above. Then enter $3 - x$ as the lower function and x as the upper function.

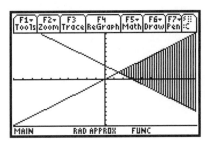

Chapter 6
Conic Sections

Many conic sections are represented by equations that are not functions. Consequently, these equations must be entered on the TI-89 as two equations, each of which is a function.

GRAPHING PARABOLAS

To graph a parabola of the form $y^2 = 4px$ or $(y-k)^2 = 4p(x-h)$, we must first solve the equation for y.

Section 6.1, Example 4 Graph the parabola $y^2 - 2y - 8x - 31 = 0$.

In the text we used the quadratic formula to solve the equation for y:

$$y = \frac{2 \pm \sqrt{32x + 128}}{2}.$$

One way to produce the graph of the parabola is to enter $y_1 = \dfrac{2 + \sqrt{32x + 128}}{2}$ and $y_2 = \dfrac{2 - \sqrt{32x + 128}}{2}$, select a window, and press $\boxed{\diamond}$ $\boxed{\text{GRAPH}}$ to see the graph. Here we use $[-16, 16, -8, 8]$. The first equation produces the top half of the parabola and the second equation produces the lower half.

 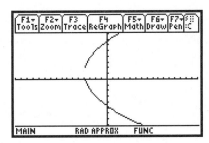

We can also enter $y_1 = \sqrt{32x + 128}$ and then enter $y_2 = \dfrac{2 + y_1(x)}{2}$ and $y_3 = \dfrac{2 - y_1(x)}{2}$. For example, to enter $y_2 = \dfrac{2 + y_1(x)}{2}$ position the cursor beside "$y_2 =$" and press $\boxed{(}$ $\boxed{2}$ $\boxed{+}$ $\boxed{\text{Y}}$ $\boxed{1}$ $\boxed{(}$ $\boxed{\text{X}}$ $\boxed{)}$ $\boxed{)}$ $\boxed{\div}$ $\boxed{2}$. Enter $y_3 = \dfrac{2 - y_1}{2}$ in a similar manner. Finally, deselect y_1 by highlighting the expression for y_1 and pressing $\boxed{\text{F4}}$. Note that there is no longer a check mark beside y_1. This indicates that this equation has been deselected and thus its graph will not appear with the graph of the equations that remain selected. The top half of the graph is produced by y_2 and the lower half by y_3. The expression for y_1 was entered to avoid entering the square root more than once. By deselecting y_1 we prevent its graph from appearing on the screen with the graph of the parabola.

We could also use the standard equation of the parabola found in the text:

$$(y - 1)^2 = 8(x + 4).$$

Solve this equation for y.

$$y - 1 = \pm\sqrt{8(x + 4)}$$
$$y = 1 \pm \sqrt{8(x + 4)}$$

Then enter $y_1 = 1 + \sqrt{8(x + 4)}$ and $y_2 = 1 - \sqrt{8(x + 4)}$, or enter $y_1 = \sqrt{8(x + 4)}$, $y_2 = 1 + y_1(x)$, and $y_3 = 1 - y_1(x)$, and deselect y_1 as described above.

GRAPHING CIRCLES

The equation of an ellipse must be solved for y before it can be entered on the TI-89.

Section 6.2, Example 1 Graph the circle $x^2 + y^2 - 16x + 14y + 32 = 0$.

In the text we found the standard form for the equation of the circle and then solved for y:

$$y = -7 \pm \sqrt{81 - (x - 8)^2}.$$

We could also have solved the original equation using the quadratic formula.

One way to produce the graph is to enter $y_1 = -7 + \sqrt{81 - (x - 8)^2}$ and $y_2 = -7 - \sqrt{81 - (x - 8)^2}$, select a square window, and press $\boxed{\diamond}$ $\boxed{\text{GRAPH}}$. Here we use $[-16, 28, -18, 4]$. The first equation produces the top half of the circle and the second equation produces the lower half.

We can also enter $y_1 = \sqrt{81 - (x - 8)^2}$ and then enter $y_2 = -7 + y_1(x)$ and $y_3 = -7 - y_1(x)$. Then deselect y_1, select a square window, and press $\boxed{\diamond}$ $\boxed{\text{GRAPH}}$. We use y_1 to eliminate the need to enter the square root more than once. Deselecting it prevents the graph of y_1 from appearing on the screen with the graph of the circle. The top half of the graph is produced by y_2 and the lower half by y_3.

GRAPHING ELLIPSES

The equation of an ellipse must be solved for y before it can be entered on a TI-89. The procedure for graphing an ellipse of the form $\dfrac{x^2}{a^2} + \dfrac{y^2}{b^2} = 1$ or $\dfrac{x^2}{b^2} + \dfrac{y^2}{a^2} = 1$ is described on page 536 of the text. Here we consider ellipses of the form $\dfrac{(x-h)^2}{a^2} + \dfrac{(y-k)^2}{b^2} = 1$ or $\dfrac{(x-h)^2}{b^2} + \dfrac{(y-k)^2}{a^2} = 1$

Section 6.2, Example 4 Graph the ellipse $4x^2 + y^2 + 24x - 2y + 21 = 0$.

Completing the square in the text, we found that the equation can be written as

$$\frac{(x+3)^2}{4} + \frac{(y-1)^2}{16} = 1.$$

Solve this equation for y.

$$\frac{(x+3)^2}{4} + \frac{(y-1)^2}{16} = 1$$

$$\frac{(y-1)^2}{16} = 1 - \frac{(x+3)^2}{4}$$

$$(y-1)^2 = 16 - 4(x+3)^2 \qquad \text{Multiplying by 16}$$

$$y - 1 = \pm\sqrt{16 - 4(x+3)^2}$$

$$y = 1 \pm \sqrt{16 - 4(x+3)^2}$$

Now we can produce the graph in either of two ways. One is to enter $y_1 = 1 + \sqrt{16 - 4(x+3)^2}$ and $y_2 = 1 - \sqrt{16 - 4(x+3)^2}$, select a square window, and press $\boxed{\diamond}$ $\boxed{\text{GRAPH}}$. Here we use $[-15, 9, -6, 6]$. The first equation produces the top half of the ellipse and the second equation produces the lower half.

We can also enter $y_1 = \sqrt{16 - 4(x+3)^2}$ and then enter $y_2 = 1 + y_1(x)$ and $y_3 = 1 - y_1(x)$. Deselect y_1, select a square window, and press $\boxed{\diamond}$ $\boxed{\text{GRAPH}}$. We use y_1 to eliminate the need to enter the square root more than once. Deselecting it prevents the graph of y_1 from appearing on the screen with the graph of the ellipse. The top half of the graph is produced by y_2 and the lower half by y_3.

We could also begin by using the quadratic formula to solve the original equation for y.

$$4x^2 + y^2 + 24x - 2y + 21 = 0$$

$$y^2 - 2y + (4x^2 + 24x + 21) = 0$$

$$y = \frac{-(-2) \pm \sqrt{(-2)^2 - 4 \cdot 1 \cdot (4x^2 + 24x + 21)}}{2 \cdot 1}$$

$$y = \frac{2 \pm \sqrt{4 - 16x^2 - 96x - 84}}{2}$$

$$y = \frac{2 \pm \sqrt{-16x^2 - 96x - 80}}{2}$$

Then enter $y_1 = \dfrac{2 + \sqrt{-16x^2 - 96x - 80}}{2}$ and $y_2 = \dfrac{2 - \sqrt{-16x^2 - 96x - 80}}{2}$, or enter $y_1 = \sqrt{-16x^2 - 96x - 80}$, $y_2 = \dfrac{2 + y_1(x)}{2}$, and $y_3 = \dfrac{2 - y_1(x)}{2}$, and deselect y_1.

Select a square window and press $\boxed{\diamond}$ $\boxed{\text{GRAPH}}$ to display the graph shown on the previous page.

GRAPHING HYPERBOLAS

As with equations of circles, parabolas, and ellipses, equations of hyperbolas must be solved for y before they can be entered on a TI-89.

Section 6.3, Example 2 Graph the hyperbola $9x^2 - 16y^2 = 144$.

First solve the equation for y.
$$9x^2 - 16y^2 = 144$$

$$-16y^2 = -9x^2 + 144$$

$$y^2 = \frac{-9x^2 + 144}{-16}$$

$$y = \pm\sqrt{\frac{-9x^2 + 144}{-16}}, \text{ or } \pm\sqrt{\frac{9x^2 - 144}{16}}$$

It is not necessary to simplify further.

Now enter $y_1 = \sqrt{\dfrac{9x^2 - 144}{16}}$ and either $y_2 = -\sqrt{\dfrac{9x^2 - 144}{16}}$ or $y_2 = -y_1(x)$, select a square window, and press $\boxed{\diamond}$ $\boxed{\text{GRAPH}}$. Here we use $[-12, 12, -6, 6]$. The top half of the graph is produced by y_1 and the lower half by y_2.

 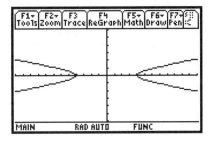

Section 6.3, Example 3 Graph the hyperbola $4y^2 - x^2 + 24y + 4x + 28 = 0$.

In the text we completed the square to get the standard form of the equation. Now solve the equation for y.

$$\frac{(y+3)^2}{1} - \frac{(x-2)^2}{4} = 1$$

$$(y+3)^2 = \frac{(x-2)^2}{4} + 1$$

$$y + 3 = \pm\sqrt{\frac{(x-2)^2}{4} + 1}$$

$$y = -3 \pm \sqrt{\frac{(x-2)^2}{4} + 1}$$

The graph can be produced in either of two ways. One is to enter $y_1 = -3 + \sqrt{\dfrac{(x-2)^2}{4} + 1}$ and $y_2 = -3 - \sqrt{\dfrac{(x-2)^2}{4} + 1}$, select a square window, and press $\boxed{\diamond}$ $\boxed{\text{GRAPH}}$. Here we use $[-15, 15, -9, 6]$. The first equation produces the top half of the hyperbola and the second the lower half.

We can also enter $y_1 = \sqrt{\dfrac{(x-2)^2}{4} + 1}$, $y_2 = -3 + y_1(x)$, and $y_3 = -3 - y_1(x)$. Then deselect y_1, select a square window, and press $\boxed{\diamond}$ $\boxed{\text{GRAPH}}$. Again, y_1 is used to eliminate the need to enter the square root more than once. Deselecting it prevents the graph of y_1 from appearing on the screen with the graph of the hyperbola. The top half of the graph is produced by y_2 and the lower half by y_3.

Chapter 7
Sequences, Series, and Combinatorics

The computational capabilities of the TI-89 can be used when working with sequences, series, and combinatorics.

FINDING THE TERMS OF A SEQUENCE

Section 7.1, Example 2 Use a graphing calculator to find the first 5 terms of the sequence whose general term is given by $a_n = n/(n + 1)$.

Although we could use a table, we will use the Seq feature from the MATH List menu. Select Auto for the Exact/Approx mode setting so that the terms will be expressed in fractional form. Press $\boxed{2nd}$ $\boxed{MATH}$ 3 to display the MATH List menu. Then press 1 or $\boxed{ENTER}$ to paste "seq(" to the entry line of the home screen. Now enter the general term of the sequence. Follow this with the variable and the numbers of the first and last terms desired. Press $\boxed{X}$ $\boxed{\div}$ $\boxed{(}$ $\boxed{X}$ $\boxed{+}$ 1 $\boxed{)}$ $\boxed{,}$ $\boxed{X}$ $\boxed{,}$ 1 $\boxed{,}$ 5 $\boxed{)}$ $\boxed{ENTER}$.

FINDING PARTIAL SUMS

We can use a TI-89 to find partial sums of a sequence when a formula for the general term is known.

Section 7.1, Example 6 Use a graphing calculator to find S_1, S_2, S_3, and S_4 for the sequence whose general term is given by $a_n = n^2 - 3$.

We will use the cumSum feature from the MATH List menu. The calculator will write the partial sums as a list. First press $\boxed{2nd}$ $\boxed{MATH}$ 3 7 to paste "cumSum(" to the home screen. Then press $\boxed{2nd}$ $\boxed{MATH}$ 3 ENTER or $\boxed{2nd}$ $\boxed{MATH}$ 3 1 to paste "seq(" into the cumSum expression. Finally press $\boxed{X}$ $\boxed{\wedge}$ 2 $\boxed{-}$ 3 $\boxed{,}$ $\boxed{X}$ $\boxed{,}$ 1 $\boxed{,}$ 4 $\boxed{)}$ $\boxed{)}$ $\boxed{ENTER}$. We see that $S_1 = -2, S_3 = -1, S_3 = 5$, and $S_4 = 18$.

Section 7.1, Example 7 (a) Evaluate $\displaystyle\sum_{k=1}^{5} k^3$.

We will use the sum feature from the MATH List menu along with the seq feature. Press $\boxed{\text{2nd}}$ $\boxed{\text{MATH}}$ 3 6 $\boxed{\text{2nd}}$ $\boxed{\text{MATH}}$ 3 $\boxed{\text{ENTER}}$ $\boxed{\text{X}}$ $\boxed{\wedge}$ 3 $\boxed{,}$ $\boxed{\text{X}}$ $\boxed{,}$ 1 $\boxed{,}$ 5 $\boxed{)}$ $\boxed{)}$ $\boxed{\text{ENTER}}$.

RECURSIVELY DEFINED SEQUENCES

Recursively defined sequences can also be entered on a TI-89 set in Seq mode.

Section 7.1, Example 9 Find the first 5 terms of the sequence defined by

$$a_1 = 5, \ a_{k+1} = 2a_k - 3, \text{ for } k \ge 1.$$

Press $\boxed{\diamond}$ $\boxed{\text{Y} =}$ and enter the recursive function by positioning the cursor beside $u_1 =$ and pressing $\boxed{\text{ENTER}}$ 2 $\boxed{\text{alpha}}$ $\boxed{\text{U}}$ 1 $\boxed{(}$ $\boxed{\text{alpha}}$ $\boxed{\text{N}}$ $\boxed{-}$ 1 $\boxed{)}$ $\boxed{-}$ 3 $\boxed{\text{ENTER}}$. Then press $\boxed{\text{2nd}}$ $\boxed{\{}$ 5 $\boxed{\text{2nd}}$ $\boxed{\}}$ $\boxed{\text{ENTER}}$ to enter the initial value of 5 beside $ui_1 =$. ({ and } are the second operations associated with the $\boxed{(}$ and $\boxed{)}$ keys, respectively.)

Next press $\boxed{\diamond}$ $\boxed{\text{TblSet}}$ to display the Table Setup screen. Set Independent to Auto, tblStart $= 1$, Δtbl $= 1$. Also press $\boxed{\diamond}$ $\boxed{\text{WINDOW}}$ and check to be sure that nmin has the same value as tblStart. Then press $\boxed{\diamond}$ $\boxed{\text{TABLE}}$ to display the table of values for the recursive function. We see that $a_1 = 5$, $a_2 = 7$, $a_3 = 11$, $a_4 = 19$, and $a_5 = 35$.

EVALUATING FACTORIALS, PERMUTATIONS, AND COMBINATIONS

Operations from the MATH Probability menu can be used to evaluate factorials, permutations, and combinations.

Section 7.5, Exercise 6 Evaluate 7!.

On the home screen press 7 $\boxed{\text{2nd}}$ $\boxed{\text{MATH}}$ 7 1 $\boxed{\text{ENTER}}$ or 7 $\boxed{\text{2nd}}$ $\boxed{\text{MATH}}$ 7 $\boxed{\text{ENTER}}$ $\boxed{\text{ENTER}}$. These keystrokes enter 7, display the MATH Probability menu, select item 1, !, from that menu, and then cause 7! to be evaluated. The result is 5040.

Section 7.5, Exercise 9 Evaluate $\dfrac{9!}{5!}$.

Press 9 $\boxed{\text{2nd}}$ $\boxed{\text{MATH}}$ 7 1 $\boxed{\div}$ 5 $\boxed{\text{2nd}}$ $\boxed{\text{MATH}}$ 7 1 $\boxed{\text{ENTER}}$. Both 1's can be replaced by $\boxed{\text{ENTER}}$ if desired. The result is 3024.

Section 7.5, Example 3 (a) Compute $_4P_4$.

Press $\boxed{\text{2nd}}$ $\boxed{\text{MATH}}$ 7 2 4 $\boxed{,}$ 4 $\boxed{)}$ $\boxed{\text{ENTER}}$. These keystrokes display the MATH Probability menu, select item 2, $_nP_r$, from that menu, enter 4 for 4 objects and 4 for 4 objects taken at a time, and then cause the calculation to be performed. The result is 24.

Section 7.5, Example 6 Compute $_8P_4$.

Press $\boxed{\text{2nd}}$ $\boxed{\text{MATH}}$ 7 2 8 $\boxed{,}$ 4 $\boxed{)}$ $\boxed{\text{ENTER}}$. These keystrokes display the MATH Probability menu, select item 2, $_nP_r$, from that menu, enter 8 for 8 objects and 4 for 4 objects taken at a time, and then cause the calculation to be performed. The result is 1680. The previous entry could also be edited to obtain this result.

Section 7.6, Example 2 Evaluate $\begin{pmatrix} 7 \\ 5 \end{pmatrix}$.

Press $\boxed{\text{2nd}}$ $\boxed{\text{MATH}}$ 7 3 7 $\boxed{,}$ 5 $\boxed{)}$ $\boxed{\text{ENTER}}$. These keystrokes display the MATH Probability menu, select item 3, $_nC_r$, from that menu, enter 7 for 7 objects and 5 for 5 objects taken at a time, and then cause the calculation to be performed. The result is 21.

Index
TI-89 Graphics Calculator